TRUE WEEVILS (PART I)

COLEOPTERA: CURCULIONIDAE
(Subfamilies RAYMONDIONYMINAE to SMICRONYCHINAE)

By
M. G. Morris

Handbooks for the
Identification of British Insects

Vol. 5, Part 17b

Editor: D. Hollis

TRUE WEEVILS (PART I)

COLEOPTERA: CURCULIONIDAE
(Subfamilies RAYMONDIONYMINAE to SMICRONYCHINAE)

By
M. G. Morris

Scientific Associate
The Natural History Museum
Cromwell Road, London SW7 5BD

(With 24 whole-insect figures by R. W. J. Read)

2002

Published by the
ROYAL ENTOMOLOGICAL SOCIETY
and the
FIELD STUDIES COUNCIL

Contents

Abstract

An introduction is given to weevils of the subfamilies Raymondionyminae to Smicronychinae, designated as part of the 'true weevils' (family Curculionidae). Difficulties occasioned by the fluid state of the higher classification of the Curculionoidea are outlined. Summaries are given of the biology, life-history and phenology, economic importance, distribution and conservation of the species covered. A checklist of species is included.

Illustrated keys are provided to the subfamilies of British Curculionidae, and to the species covered in this handbook. The entry for each species includes characters for distinguishing the sexes and information on biology, larval foodplants, distribution within the British Isles and zoogeographic range. Those species with 'Red Data Book' status are indicated.

Acknowledgements

I gratefully acknowledge the assistance given me by the institutions and individuals named in the previous handbooks (to orthocerous and broad-nosed weevils). I thank particularly the Keeper of Entomology, the Natural History Museum, and the staff of the Coleoptera Section and Entomological Library of the Museum for their ever-ready willingness to assist with enquiries and problems.

Thanks are also due to Mr M.V.L. Barclay, Dr R. Caldara, Mr T. Eccles and Dr G. Osella, in addition to those named in previous volumes.

Mr R.W. John Read is thanked once again for providing drawings of whole insects for this *Handbook*.

Financial assistance towards production of the work, provided by the British Entomological and Natural History Society, is gratefully acknowledged.

'True Weevils' (part I): Family Curculionidac (Phanerognatha), subfamilies Raymondionyminae to Smicronychinae

Introduction

When a series of *Handbooks* for the identification of British weevils was begun, the checklist of Pope (1977) was followed. It was (and is) intended to cover the superfamily Curculionoidea, excluding Scolytidae and Platypodidae (already keyed by Duffy, 1954). However, since the appearance of a *Handbook* to the Orthocerous groups (Morris, 1990) and during preparation of one to the 'broad-nosed' species (Morris, 1997) considerable changes have been proposed to the higher systematics of the Curculionoidea. Despite the thoroughness of the treatments of Thompson (1992), Kuschel (1995), Zimmermann (1993, 1994a, 1994b) and Lawrence & Newton (1995) (this last publication covering the whole of the Coleoptera, not exclusively Curculionoidea), there is still controversy over many issues which militates against an accepted system. Moreover, specialist contributions to 'total evidence', such as the fine structure of spermatozoa (Burrini *et al.,* 1988) or wing venation (Zherikhin & Gratshev, 1995) have tended to cut across the arrangements and classifications proposed by these recent authors. A particular issue is the division of the Curculionoidea into 'Orthoceri' and 'Gonatoceri'. Originally distinguished by the absence or presence of geniculate antennae respectively, the division is now based mainly on the structure of the male genitalia (e.g. Thompson, 1992). Thus some groups of weevils are 'orthocerous' even though they have geniculate antennae.

More recently still, in an authoritative catalogue of World curculionoid families and genera, Alonso-Zarazaga & Lyal (1999) have used an arrangement which may well be accepted as at least an interim standard. It has the merits of comprehensiveness (down to subtribe in many cases) and incorporates the results of recent research. However, it differs considerably from more traditional arrangements, particularly in the groups subsumed in Curculioninae *s. l.*

In deciding how to treat the 'true weevils' (the groups not covered by Duffy (1954) or Morris (1990, 1997)) there was an intention to recognise that advances had been made since the comprehensive revision of the higher systematics of Coleoptera by Crowson (1967). However, a balance needed to be struck between modernity and the understandable desire of many coleopterists for a stable, or long-lasting arrangement. One of a number of options was possible. Strict adherence to Pope (1977) was not one of these because of changes in the classification of particular genera and species. For example, *Dryophthorus* would have had to be transferred from Cossoninae to 'Rhynchophorinae' and subfamily 'Erirhininae' dismembered in recognition of its polyphyletic nature.

There would also have been difficulties in adapting to the schemes of Kuschel (1995), Lawrence & Newton (1995), or Alonso-Zarazaga & Lyal (1999), which have many similarities. All three arrangements subsume many of the subfamilies recognised by Pope (1977) into a single subfamily Curculioninae which, whatever the rights and wrongs of a strictly phylogenetic arrangement, is unwieldy in a purely practical sense. Moreover, the first two schemes include Entiminae, together with other subfamilies, in another large group, Brachycerinae, into which the 'broad-nosed' species (Morris, 1997) could not easily be fitted (as if to emphasise the lack of agreement Alonso-Zarazaga & Lyal (1999) have Brachyceridae as an 'orthocerous' family, no doubt following Thompson (1992)).

The recognition by Thompson (1992) and other authors of separate family status for Raymondionymidae, Dryophthoridae (amended from Rhynchophoridae) and Erirhinidae (*s. l.*), also created problems, which could only partly be solved by adopting 'divisions of convenience'(Zimmermann, 1994a). These families are 'orthocerous' and would have been included with the families treated under that title (Morris, 1990) had the information been available at the time.

As a compromise solution, the arrangement adopted here conforms in general to the arrangement of Thompson (1992), but accords only subfamily status to the three 'new' curculionid families named above. However, attention is drawn to their anomalous status in the text. Another change from Pope (1977) is that subfamilies 'Hylobiinae', 'Rhyparosominae' and 'Pissodinae' are subsumed into Molytinae, following the arrangement of Kuschel (1987, and unpublished information) and Alonso-Zarazaga & Lyal (1999).

It must be made clear at the outset that subfamily Erirhininae *sensu* Pope (1977) has been shown by Thompson (1992) to be polyphyletic and, as stated above, to contain both orthocerous and gonatocerous representatives. Coleopterists familiar with Erirhininae *sensu* Pope will find some of the genera included in the (orthocerous) Erirhininae (Erirhinidae *sensu* Thompson) while others are placed in gonatocerous subfamilies of Curculionidae (Tanysphyrinae, Bagoinae, Storeinae, Styphlinae and Smicronychinae).

Morris (1995a), drawing on the work of Thompson, Kuschel and others, set out an arrangement of British weevils, to which the present *Handbook* broadly adheres. However, the following changes were made by Morris & Booth (1997): subfamily Rhyparosominae became Molytinae-Phryxinini (Kuschel, 1964, 1987) and the polyphyletic subfamily 'Styphlinae' was divided into Pachytychiinae, 'Dorytominae' and Styphlinae *s. str.* Some of these groups are controversial, with a particular problem attached to the genus *Dorytomus*, the taxonomic position of which was in doubt. Alonso-Zarazaga & Lyal (1999) placed *Dorytomus* in Curculioninae-Ellescini (as subtribe Dorytomina), and placed other groups covered here also in Curculioninae.

Tanysphyrus is gonatocerous, as noted above, and is placed in subfamily Tanysphyrinae pending further work on its status. Its gonatocerous nature was discovered too late for it to be correctly positioned in the arrangement of Alonso-Zarazaga & Lyal (1999).

Admittedly, there is a proliferation of subfamilies in the present *Handbook* and this runs counter to the desire of many weevil specialists to reduce their number. Some better use of infrasubfamily categories may be the solution in the long term, and this is a feature of the arrangement of Alonso-Zarazaga & Lyal (1999).

This *Handbook*, then, covers Raymondionyminae, Dryophthorinae and Erirhininae (properly families) and the curculionid subfamilies Lixinae to Smicronychinae. These groups equate to Cleoninae to Erirhininae of Pope's checklist (1977).

Species of *Brachycerus* (Brachyceridae *sensu* Thompson, 1992) are occasionally imported into Britain in garlic and other bulbs, but are not considered to be part of the British fauna. In passing, it should be noted that subfamily Brachycerinae *sensu* Kuschel (1995) and *sensu* Lawrence & Newton (1995) subsumes a large number of curculionid groups, including the broad-nosed species (Entiminae). European authors generally (e.g. Abbazzi & Osella, 1992) have adopted a much more restricted concept of Brachycerinae, including only *Brachycerus*. Thompson (1992), followed by Alonso-Zarazaga & Lyal (1999), showed that *Brachycerus* and its allies are orthocerous, whilst the Entiminae (dealt with by Morris, 1997) are part of Curculionidae.

Biology

The biological characteristics of the groups of weevils dealt with here are exceptionally diverse. Although, like the great preponderance of weevils, all the species feed on plants, within this constraint almost every kind of exploitative strategy is adopted by one or other of the groups covered. To some extent the pattern of this diversity is confused by the classification adopted here. In more recent arrangements the subfamily Curculioninae in its extended sense (Anderson, 1995; Lawrence & Newton, 1995; Alonso-Zarazaga & Lyal, 1999) subsumes most, though by no means all, of the stenophagous species which feed on living plants. References to the biological characteristics of the different species mentioned below are given in the systematic species accounts (keys).

The recent discovery of *Ferreria* (*Raymondionymus* auctt.) in Britain has increased the diversity of our weevil fauna. *Ferreria marqueti* is the only blind, eyeless species to occur in this country. It is almost certainly a root-feeder, at considerable depth in the soil, and normally passes its entire life below ground, as its eyeless state indicates.

Not many other species of those covered here are root-feeders, though several feed in the lower stems or rootstocks (*Liparus* spp., *Mitoplinthus caliginosus*), and in rhizomes (e.g. *Leiosoma deflexum*).

Dead wood is the larval substrate of a number of groups. All British Cossoninae inhabit dead wood as larvae and adults, with both stages being found at all seasons of the year. Adults of some cossonines are also found away from dead wood, but this is unusual. Apparent preferences of some species for certain species of trees have been disproved by later work. Thus *Rhopalomesites tardyi* ('Holly weevil'), supposedly with a preference for *Ilex aquifolium*, actually infests dead wood of a wide variety of mostly broad-leaved tree species. *Pselactus spadix* was formerly thought to infest primarily wood of deciduous trees in maritime situations (groynes, wharves and driftwood) but is now known to prefer wood of conifers. However, most of our cossonines are found mainly in the wood of broad-leaved trees. *Cossonus* spp. are often found in dead wood of *Salix* and *Populus* spp., and also occasionally in the dead stems of other plants, such as cabbages. *Rhyncolus ater* is exceptional in being usually an inhabitant of *Pinus* spp.; it is often found under bark of stumps of *P. sylvatica*. *Dryophthorus corticalis*, now placed in a distinct family, but formerly regarded as cossonine, is also xylophagous, being found especially in dead wood of *Quercus*.

British species of *Acalles* probably feed in dead wood of trees, and of *Calluna vulgaris* in the case of *A. ptinoides*, but their biology is not well-known. *Trachodes hispidus* is another little-known species which may feed in dead wood; adults are often found in bundles of faggots, but also elsewhere. *Anchonidium unguiculare* occurs among woodland litter at some sites but, as it also occurs in unwooded situations, it is unlikely to be dependent on dead wood.

The biology of *Magdalis* species, another major group of wood-feeding weevils, is very different from that of cossonines. Larvae are subcortical feeders in a variety of trees, but the weevils are stenophagous or oligophagous. Several species occur on *Pinus*, others on Rosaceae, and two species attack *Ulmus* and *Betula* respectively. Unlike cossonines, adults have a short life, the weevils having a long larval stage, in which they overwinter. The adults are much more commonly found on the foliage of their hosts, or even away from

7

them. Whereas cossonine adults lay their eggs while in galleries in dead wood, female *Magdalis* oviposit from without, and have been observed drilling a hole with the rostrum and inserting an egg into the preferred substrate with it.

Of greater significance economically are the species of *Pissodes* and *Hylobius* which feed in dead wood of conifers and also damage living trees; they are considered under the species of economic importance.

The stems of woody plants are the larval feeding site of only a few species, most notably *Cryptorhynchus lapathi*, which is also a species of economic significance, on osiers. Stems of *Equisetum* spp. are utilised by *Grypus equiseti*, while *Lixus* spp. and cleonines feed as larvae in the stems of various large herbs. The introduced species *Syagrius intrudens* feeds in the stems of ferns. Stems of monocotyledons are probably the larval feeding sites of many of our Erirhininae, but the biology of these species is not well-known in Britain.

Foliage-feeding weevils (of those dealt with here) fall into two groups, those which mine leaves and those which feed ectophagously. Stenophagous leaf-miners include *Anoplus plantaris* (on *Betula*), *A. roboris* (on *Alnus*), and *Bagous alismatis* (on *Alisma plantago-aquatica*). *Orthochaetes* spp. are much more varied in their choice of hosts, both British species apparently mining a very wide range of both monocotyledenous and dicotylenonous plants, though the range is not well-known in Britain.

Two main groups of ectophagous species are dealt with here, Cioninae and Hyperinae, though larvae of both *Stenopelmus rufinasus* and *Tanysphyrus lemnae* feed semi-exposed on their hosts. *Cionus* and *Cleopus* feed mainly on Scrophulariaceae, though also on garden Buddlejaceae; their slug-like larvae are familiar objects, as are the pupal cocoons. Unlike the larvae, which are conspicuous and presumably distasteful to potential predators, the cocoons are cryptic, often hidden among the inflorescences of the hosts. Adult *Cionus* are also cryptic, with patterns that mimic seed capsules. *Hypera* and *Limobius* larvae are caterpillar-like and, unusually for weevil larvae, pigmented. The pupal cocoons are characteristic and briefly described for each species, for which they are known, in the systematic section. Species of Hyperinae feed either on one species of plant in Britain (e.g. *H. pastinaceae* on *Daucus carota*, *H. plantaginis* on *Lotus corniculatus*) or plants within one family (e.g. *H. arator* on Caryophyllaceae, *H. pollux* on Apiaceae).

The only gall-inducers among the weevils covered here are the two species of *Smicronyx* which induce stem-galls on *Cuscuta* and the rare *Bothynoderes affinis* which induces galls on the roots of Chenopodiaceae. The unpredictable occurrence of the host dodders (*Cuscuta*) is reflected in the rarity of *Smicronyx*, especially *S. coecus*, although as the hosts of the common *C. epithymum* are *Ulex, Cytisus* and *Calluna* spp. the weevils are found usually in association with those plants.

Feeders in inflorescences and in fruits and seeds are not well-represented among the species dealt with here. Notable among the inflorescence-feeders are *Larinus planus* and *Rhinocyllys conicus* in 'thistle' capitula and *Pseudostyphlus pillumus* in those of mayweeds. *Pachytychius haematocephalus* feeds as a larva in the pods of *Lotus corniculatus*. The catkins of Salicaceae appear to be the only ones suitable for weevil larvae, those of *Quercus* and *Corylus* being probably too dry. Nearly all *Dorytomus* spp. have larvae which feed in catkins, though some appear to utilise vegetative buds or shoots, at least facultatively. Different patterns of timing and resource partitioning have been demonstrated for some species associated with the same salicaceous tree (Morris, 1998).

A number of the species covered here are aquatic or semi-aquatic. They attack many different water plants, but the feeding biology of many, most notably *Bagous* species, is very poorly understood. The hosts of *Stenopelmus rufinasus* (the water fern, *Azolla*) and *Tanysphyrus lemnae* (duckweeds, *Lemna* spp.) are well-known, as are some of the putative hosts, if not the feeding biology, of some *Bagous* spp. (e.g. *B. limosus* on *Potomageton, B. frit* on *Menyanthes*).

Some of these aquatic species are well-adapted to life in water, whereas others have few adaptations to an existence in ponds or streams. None is able to live in fast-flowing water. *Tanysphyrus lemnae* and *Stenopelmus rufinasus* are unable to swim, but some *Bagous* have swimming ability and have efficient plastron respiration. However, their pattern of underwater existence is based on their ability to climb and crawl on submerged vegetation; the well-developed tibial hooks or mucrones are evidence for this. A few species have the ability to skate or swim on the surface of water, though this power is much better developed in some Ceutorhynchinae. *Hypera rumicis* is, surprisingly, one of these, but no skating ability was found in preliminary tests on *Notaris scirpi, Cionus tuberculosus* or *Grypus equiseti*, which might have been expected to be better adapted to an aquatic existence than *H. rumicis*.

Flight is well-developed in most of the species discussed here, though most of the bulkier Molytinae are flightless. However, locomotion and dispersal have been little-studied in most species. Both British *Orthochaetes* species are flightless, as is *Sitophilus granarius*, though the other two species in the genus are fully winged and fly readily. Most of those Cossoninae which have been studied, even those with well-developed wings, fly rarely, if at all.

As with other groups of weevils, different species are found at ground level, on herbaceous vegetation and on trees, usually in relation to their host plants. There is a good, but by no means perfect, correspondence between size and living on the ground. Our largest and bulkiest species of *Hypera, H. punctata*, is the species most often found on the ground, whereas most of the other, smaller, species occur on field layer vegetation. *Liparus* spp. and *Mitoplinthus caliginosus* are ground-living. But there are also much smaller species with the same habit, for example *Leiosoma, Gronops* and *Orthochaetes* spp. and *Anchonidium unguiculare*.

Parthenogenesis is known in very few long-nosed weevils compared with their broad-nosed counterparts (Entiminae). Both British *Orthochaetes* species are parthenogenetic. Most of the other species treated here seem to have 50:50 sex ratios. Secondary sexual dimorphism is strongly developed in a few species, notably *Rhopalomesites tardyi* (Cossoninae) and some subgenera of *Magdalis*. In others, for example many *Bagous* spp., there is little external difference between the sexes. However, the males of many species have better developed tibial or femoral teeth, associated presumably with the holding of females during copulation.

Life-history and phenology

Most weevils conform to a simple pattern of life-history: there is a single generation a year, adults overwinter in reproductive diapause, mating occurs either in spring or the previous autumn, and the immature stages are passed in late spring and summer with adults of the year appearing in late summer and autumn.

However, there are many exceptions to this pattern. Some of those species which feed on plant structures that are available over a long period, such as foliage, have more than one generation a year. *Cleopus pulchellus*, feeding on foliage of *Scrophularia* spp., has two generations a year (Read, 1976), although species of *Cionus*, with similar biology, have only one (Read, 1977). *Hypera punctata*, developing on *Trifolium* foliage, appears to have two independent generations a year (Scherf, 1964). In California, *Steopelmus rufinasus* probably has 4-6 generations a year and completes its life-cycle in 11 days at constant 90°F in the laboratory (Richerson & Grigarick, 1967). There is no information on the usual number of annual generations in Britain. The influence of temperature on the number of generations produced by British weevils generally has been little studied.

Where larval food is available all the year round species may be continuously brooded, the duration of the life-cycle being temperature- (and humidity-) dependent and strongly influenced by the nutritive value of the food. Such is probably the case with Cossoninae and other species living continuously in dead wood. *Pentarthrum huttoni* was reared by Hammad (1955); the life-cycle took 4-4.5 months at constant 25°C and 95% RH. The slow rate of development probably reflects the poor nutritive value of dead wood. *Sitophilus* spp. are notoriously continuously brooded and their rates of development on the highly nutritive grain are much more rapid.

Species of *Magdalis*, which feed in dead wood, but in simple galleries under the bark of dead, and possibly dying or unhealthy, branches have a different type of life-history. The species are not continuously brooded but have a life-cycle which is similar to those of most species feeding on plant structures, but with modifications to the duration of the different stages. The adult stage is short-lived, with the larval stage the longest, unlike foliage-feeding and most other weevils but similar to many root-feeding Entiminae (Morris, 1997). Again, this almost certainly reflects differences in the nutritive value of the larval substrates; *Magdalis* larvae need a long feeding period to develop. In other wood-living species the larval stage may be even longer. *Hylobius abietis* takes 1-2 years to develop in Britain and three years or longer further north in Europe (Bevan, 1987). Species of *Pissodes* have similar life-history patterns to those of *Magdalis* spp., but it is interesting to find that *P. validirostris* (which feeds as a larva in pine cones) apparently has a similar life-history to the two other British species, which feed in dead branches of *Pinus*.

For weevils dependent on short-lived plant structures, such as buds, fruits and seeds, modification of the 'normal' pattern of life-history is in its timing. Most of the plant structures on or in which larvae feed are available only in summer. Some obvious examples are the rootstocks, stems and flower-heads of annual or perennial herbs in which species of Lixinae and some Molytinae feed. Seed-feeding species include *Pachytychius haematocephalus* (on *Lotus corniculatus*) and *Smicronyx reichi* (on *Centaurium erythraea*); fruits of both hosts are produced in summer. However, there are exceptions, some of the clearest being the catkins of Salicaceae in which larvae of *Dorytomus* spp. feed. Catkins of both *Salix* and *Populus* species appear in very early spring, and the life-histories of *Dorytomus* are modified accordingly. In the species feeding on *Populus tremula*, adults are active and oviposit either in late autumn (*D. tortrix*), or late winter to early spring (*D. affinis* and *D. dejeani*) (Morris, 1998). In these species the resting period

10

of adults is long and qualifies as aestivation rather than hibernation. Similar modifications are probably to be found in the *Dorytomus* spp. feeding on other *Populus* and *Salix* spp., but have been little-studied.

It is not known what effects the variable and unpredictable appearance of host plants has on the life-history and phenological characteristics of the weevils dependent on them. It is probable that the success of species such as *Smicronyx jungermanniae* depends on effective dispersal and on detection (presumably chemical) of the species of *Cuscuta* on which it induces gall formation.

There is also little information on whether the adults of some species can survive and breed for more than one year, and whether variation in weather can produce different patterns of phenology in different years. Regional variation in life-history patterns has also been little-studied in weevils. Long-term climate change may produce more fundamental and long-lasting effects. However, more accurate information is required before work on weevils can contribute very much in this area.

Economic importance

The weevils dealt with here include some major pests and many more of lesser significance. None is deleterious to human health, of course, though it is said anecdotally that *Lixus paraplecticus*, when it was commoner in England than it has been in recent decades, occasionally caused injury to grazing cattle which inadvertently ingested adults (Stephens, 1831).

The major areas of economic activity, agriculture, horticulture, forestry, building and the storage of food products, are all subject to attack by weevils.

'Clover leaf weevils' (*Hypera postica* and *H. nigrirostris*) are minor pests of clover grown for seed (Gratwick, 1992), affecting yield only in very dry seasons (though *H. postica* is more generally associated with *Medicago* spp.). Larvae of *Mitoplinthus caliginosus* can cause serious damage to hop bines by weakening them, and killing them in severe attacks (Alford, 1984); the adults are of little economic significance. Species of *Magdalis* are very minor pests of several top fruit trees (Massee, 1954; Alford, 1984), but are more prevalent in neglected orchards than well-managed ones.

Although many species of weevil feed on broad-leaved forest and other trees, only a few cause economic damage. *Cryptorhynchus lapathi* is a significant pest of *Salix* spp., particularly osiers (Smith & Stott, 1964). Much more important economically are the weevils associated with conifers. *Pissodes pini* and *P. validirostris* are of minor significance, but *P. castaneus* is an important pest of the young and pole stage trees of *Pinus* (Bevan, 1987). However, the most damaging pest of pines in Britain is *Hylobius abietis*, which devastates transplants in particular (Bevan, 1987).

Internationally, the most notorious weevil pests are grain weevils (*Sitophilus* spp.). Zimmerman (1993) gave a graphic account of the ravages of *S. oryzae* in Australia before control measures were introduced. The pest was particularly damaging because of its powers of dispersal by flight. The flightless *S. granarius* and fully-winged *S. zeamais* are also, or have been, important pests. In Britain at the present time all species of *Sitophilus* are uncommon, improved food hygiene being mainly responsible. The only other stored products pest to be considered here is *Caulophilus oryzae*, a rare species, mainly introduced in ship-borne cargo and associated principally with stored ginger.

Although of no significance as far as the British fauna is concerned, it can be noted that species of *Lixus, Larinus* and *Rhinocyllus* have been used in attempts for the biological control of thistles in various parts of the world (Zwölfer, 1965; Zwölfer & Harris, 1984).

Several species of cossonine are minor pests of structural timber in buildings. Hum, Glaser & Edwards (1980), in wide-ranging surveys of buildings, recognised *Pentarthrum huttoni, Euophryum confine* and *E. rufum* as sufficiently common to constitute an economic problem. *Pseudophloeophagus aeneopiceus* (as *Caulotrupodes*) was rare in buildings. *Pselactus spadix* has recently been considered to cause significant structural damage to wharf timbers in southern England (Sawyer & Cragg, 1995).

Distribution

The species dealt with here demonstrate most of the common patterns of distribution within both the British Isles and western Europe more generally. Some caution must be adopted in assessing distributions, particularly in the less well-recorded areas of the British Isles, because of inadequacies of investigation and, in some cases, of correct identification. Only one species is endemic to our area (*Procas granulicollis*), all others being found elsewhere in Europe, where they are often common and widely distributed. Detailed zoogeographical analysis of distribution patterns of British weevils remains to be made. It must be noted also that there is a constant flux of species, with distributions being highly dynamic. The examples of *Stenopelmus rufinasus, Lixus scabricollis, Euophryum* spp., *Syagrius intrudens* and *Gronops inaequalis* have to be set against several extinctions, though accidental introduction by Man is responsible for most, if not all, of these.

Rather few species are found throughout the British Isles, including the northern and western Isles of Scotland; *Notaris acridulus, Leiosoma deflexum, Hypera plantaginis* and *H. nigrirostris* are examples, and several other *Hypera* species are nearly as widely distributed. Rapid spread to inhabit most of the British Isles has been characteristic of *Euophryum confine*, originating from New Zealand and first recorded in 1937 (Hickin, 1968). Widely distributed arboreal species are usually absent from the Hebrides, Orkney and Shetland, where trees are scarce or absent. Examples are *Hylobius abietis, Anoplus plantaris* and *Dorytomus taeniatus*.

More species of weevils occur in southern England than in northern Scotland, there being a more-or-less clear gradient of diversity associated with latitude (Morris, 1999). Some species can be quite common in the extreme south of England, but extend to only varying degrees further north. *Smicronyx jungermanniae* is distributed from W Cornwall to Surrey, but extends only to Berkshire, with a more outlying occurrence in E Norfolk. *Rhinocyllus conicus* is common in some southern coastal counties, such as Dorset, but is not found elsewhere, while *Larinus planus* has only a slightly wider distribution, being found as far north as Merioneth, but with no records for the vice-counties between Surrey and W Gloucester, the northernmost English locality. *Magdalis barbicornis* has not been recorded north of Leicester, *M. cerasi* not further north than Mid-W York, and *Gronops lunatus* not further north than SE York.

Some species are distributed throughout England and Wales but extend only a short distance into southern Scotland. Examples (with their northernmost records) are *Cleonis pigra* (Angus), *Hypera dauci* and *Magdalis armigera* (W Perth; old record for the latter species), *M. ruficornis* (Stirling), *Phloeophagus lignarius* (Dumfries; old record) and *Acalles roboris* (Edinburgh).

Several species are predominantly northern in distribution, though only *Pissodes validirostris* is not known outside Scotland. However, *Magdalis duplicata* and *M. plegmatica* penetrate only a short distance from Scotland into northern England. *Erirhinus aethiops* also occurs in northern England and Scotland, but occurs in Ireland (mainly the north) as well.

The Irish fauna is impoverished compared with that of Great Britain, but general patterns of occurrence are not readily apparent. *Dorytomus tortrix* is quite widely distributed in Ireland, and not uncommon, but its Aspen-feeding congeners *D. dejeani* and *D. tremulae* (the latter admittedly a much less common species) are absent. *Cionus hortulanus* is common throughout Ireland (though it has not been recorded from Scotland), but the other common British cionines are either very rare or have not been found in Ireland. Of the species associated with *Pinus* only *Hylobius abietis* has recolonised Ireland since reafforestation of the island, species of *Pissodes* and *Magdalis* (*s. str.*) being absent. Because recording in Ireland has been much less thorough than in England, uncertainties must remain as to the exact composition of the fauna. More than 20% of the Irish weevil fauna has been discovered since 1902 (Morris, 1993a).

Only a few species have wide, but very patchy, distributions. One example is *Anoplus roboris*, which occurs throughout the British Isles as far north as Easterness. *Hypera diversipunctata* is rare and much less widely distributed, but has been recorded sporadically from W Suffolk and Radnor northwards to Edinburgh. Its flightlessness probably contributes to its very patchy distribution.

On the other hand, the weevils covered here include a large proportion of species which are very rare or very restricted in their British distributions. Species of Bagoinae contribute extensively to this group and this is characteristic of the subfamily throughout Europe. *Bagous czwalinae* has not been found outside the New Forest, while *B. brevis* has been reliably recorded from only that area, Surrey and Co. Clare. At least three bagoines are almost certainly extinct in the British Isles, and each was recorded from only a single locality, as far as is known. There are several other examples of species which have become extinct during the 20th century, particularly in the genus *Lixus*.

Some species, including several *Bagous*, are known to be British on a very few specimens, often of considerable antiquity. On the other hand, the recent discovery of *B. frit*, thought to be restricted to the New Forest, at several places in Wales shows that perceptions of distributions are heavily dependent on adequate recording activity.

Other examples of species which are extremely localised include *Hypera pastinacae*, not reliably recorded outside a small area of E Kent, *Dryophthorus corticalis*, known only from Windsor, Berkshire, and *Liparus germanus*, restricted to Kent, E Sussex and (possibly) the eastern edge of Surrey.

Most British weevils are widely distributed in western and central Europe. The patchy distributions of many *Bagous* spp., and endemic nature of *Procas granulicollis* have been noted. *Orthochaetes insignis* is restricted to the western edge of Europe and North Africa, but most of the other species which occur only sporadically in central and northern Europe have been introduced. There are no counterparts to the broad-nosed genera *Cathormiocerus* and *Caenopsis* (Morris, 1997) among those considered here.

Conservation

Aspects of wildlife conservation applicable to weevils include research, site safeguard and management, the maintenance of biodiversity, the preservation of populations of species of particular importance, the designation of species for use in decision-making, legislation and public attitudes and concern. Only superficial coverage of a few of these aspects is appropriate here.

The United Kingdom is fortunate in that a good understanding exists between conservationists on the one hand and research workers and field entomologists on the other. It is obvious that the aspirations of both groups are not only compatible but complementary.

It is also obvious that protection of sites and their management are the chief measures which will preserve all but the rarest and most important of our species. Many species of weevils probably entered the British Isles, or became commoner, because of the development of agriculture. This component is less prominent in many other groups and this has influenced attitudes to their conservation.

Species Action Plans (SAPs) are in place for a number of invertebrates. These animals have received enhanced publicity following better recognition of threats to our rarest species. The only species included here which has been singled out for such treatment is *Pachytychius haematocephalus*. Although some of the assumptions and proposed actions made in SAPs could be considered somewhat naive, the plans are to be welcomed in giving greater prominence to the problems posed by conservation of individual species, rather than sites and biotopes, and to the importance of invertebrates. *P. haematocephalus* currently exists in only a short section of the Hampshire coast, where it has been known for about 150 years. The site is under threat from development and other activities, so treatment of *P. haematocephalus* as a SAP species is appropriate.

None of the species considered here is subject to 'protection' under current legislation. However, considerable use has been made of the designations made under IUCN 'Red Data Book' categories (Shirt, 1987) and 'Nationally Notable' ones (Hyman and Parsons, 1992). The former, though not the latter, are included here, as in Morris (1997). Considerable anomalies exist in both sets of assessments, chiefly because of ongoing research and changes in the presence and abundance of species in the British Isles, many changes of course being Man-induced. Incomplete information, misidentifications and adherence to rigid criteria have also produced anomalies, though it is arguable that such criteria have some advantages over subjective assessments. The 'Extinct' category is one which is particularly difficult to assess. Ten species of those covered by this *Handbook* are considered to be extinct in the British Isles (though only seven are included by Hyman & Parsons, 1992), but the number is probably higher because only species for which there is no record after 1900 fall into this category.

The proportions of Red Data Book and Nationally Notable species are about similar for the weevils as for Coleoptera as a whole, with perhaps rather more RDB1 (Endangered)

14

species (Hyman & Parsons, 1992: Table 1). Lixinae and Bagoinae are the groups in which most RDB species fall. In many cases species are designated which have no currently known population or site of occurrence, or which are known to be British on very few specimens. These and other factors show that the perception that the British fauna is well-known is only relative. The species accounts in this *Handbook* include brief descriptions of current status; a comprehensive summary of these is inappropriate here. Designations for species which have been accidentally introduced are arguably inconsistent in Hyman & Parsons (1992), as shown by the treatment of *Lixus scabricollis* (RDB) and *Magdalis memnonia* (no designation), for example.

It is remarkable that the pest species *Mitoplinthus caliginosus* and *Cryptorhynchus lapathi* also have Nationally Notable status (Na and Nb respectively).

International aspects of conservation have received little attention, except in the case of endemic species. *Procas granulicollis* is the only species which is not recognised elsewhere; it has become better known in the British Isles in recent years, though little information is available on its biology. Although there is no group of species with extremely restricted European distribution comparable to *Cathormiocerus* (Morris, 1997), *Anchonidium unguiculare* and *Orthochaetes insignis*, for example, have very restricted ranges in the extreme west of Europe. Attention is drawn to the particular importance of *Syagrius intrudens*, which, despite belonging to an Australian group, is not known outside the British Isles.

Glossary

The terms defined below are mostly morphological ones used in the keys. However, some other terms are included if they are thought to be unusual or particularly specialised. Words of Latin or Greek origin which are used with their original plurals or singulars are indicated ('pl.', 'sing.'), as are the antonyms of some terms ('opp.').

acuminate	pointed, needle-shaped
-ad	a suffix indicating 'towards', 'in the direction of'; e.g. basad, towards the base
adelognathous	with hidden mandibles; applied to broad-nosed weevils (Entiminae)
aestivation	a resting stage in summer (cf. hibernation)
alary	pertaining to wings and flight
allopatric	taxa (*q.v.*) native to, or originating in, different regions (opp. sympatric)
anal segment	the last abdominal segment
antennal insertion	the site of articulation of the scape (*q.v.*) on the rostrum (*q.v.*)
anteriad	towards the anterior (front)
apicad	towards the apex
apical	at the apex of
apodeme	the two proximal (*q.v.*) processes of the male median lobe (*q.v.*) (used in a restricted sense)
appendiculate	of tarsal claws etc.; with an 'appendix' or internal process (e.g. figs 30, 32)
appressed	closely pressed down

arboreal	inhabiting the canopy of trees
asperity	a small rough process or area
auctt.	(auctorum) 'of authors'; used of a name which has been incorrectly applied by more than one author
auctt. Brit.	of British authors
basad	towards the base
base	the posterior part of the pronotum (*q.v.*), but (confusingly) the anterior part of the elytra (*q.v.*)
bicuspate	having two cusps or blunt points
bifid	cleft in two
bifurcate	having two branches

capitulum (pl. capitula) the inflorescence (flower head) of species of Asteraceae

carina (pl. carinae) a narrow raised ridge

carinate	with a narrow raised ridge (carina)
clavate	club-shaped
club	the distal (*q.v.*) three segments of the antenna which usually form a distinct structure
collar	the anterior part of the pronotum (*q.v.*) in some species, marked off by a subapical impression
concolorous	of the same colour
confluent	running together
connate	(of tarsal claws etc.) joined at the base
cordiform	shaped like a conventional heart
cultivar	a garden form of a plant
declivity	the distal part of the elytra (*q.v.*), where they more or less suddenly slope downwards
deflexed	bent
derm	integument, 'skin'
disc	the central part of the pronotum (*q.v.*) or elytra (*q.v.*)
discal	pertaining to the disc (*q.v.*)
distad	towards the extremity
distal	far from the body; particularly used of elytra (*q.v.*), legs and rostrum (*q.v.*)
dorsum	the upper surface
ectophagous	feeding externally
effaced	'rubbed out'; indistinct
elongate	longer than broad

elytra (sing. elytron) the wing-cases (modified fore-wings) of Coleoptera

elytral	pertaining to the elytra (*q.v.*)
emarginate	incised (*q.v.*), having an indentation
entire	complete, not evanescent (*q.v.*)
epigean	above ground

epimeron (pl. epimera) the side of the thorax above the coxae (articulation of the legs)

evanescent	fading away

excavate	scooped out
excised	cut out
exiguous	scanty, slender
explanate	formed into a narrow plate
facultative	having the facility to, not obligatory
flagellum	the antennal segments excluding the scape (*q.v.*), i.e. the funiculus and club
form	an infraspecific (*q.v.*) variant of more than casual or occasional occurrence
fossa	a cavity or depression
fugitive	liable to disappearance, transient
funiculus	the antennal segments not including the scape (*q.v.*) or club (*q.v.*)
geniculate	'kneed'; of antennae with a distinct angle between the scape (*q.v.*) and funiculus (*q.v.*) (opp. orthocerous (*q.v.*))
glabrous	bare, without setae (*q.v.*)
Gonatoceri	a division of Curculionoidea, originally those with geniculate (*q.v.*) antennae
gonatocerous	with geniculate (*q.v.*) antennae (but currently used to denote membership of the Gonatoceri (*q.v.*))
granulate	covered with granules (small, regular, raised areas)
herb	a herbaceous (non-woody) plant (not in the sense of a culinary herb)
humeral	pertaining to the humeri or humerus (*q.v.*)
humerus (pl. humeri) the shoulder of the elytra (*q.v.*); the antero-lateral angle of an elytron
hydrofuge	water-repelling
hypogean	soil-inhabiting
ICZN	The International Commission on Zoological Nomenclature
incised	cut into
infraspecific	a category below the species level
interspace	same as interstice (*q.v.*); also used of the space between punctures on other parts of the body
interstice	the space between elytral (*q.v.*) striae (*q.v.*)
interstice 1, etc.	the interstices are numbered outwards from the suture (*q.v.*)
isodiametric	of punctures etc.; as long as wide
laterad	towards the side
macula (pl. maculae)a small marking, often formed by a patch of pubescence or scales (*qq.v.*)
median lobe	the male intromittent organ, or penis
-mere	a segment or joint; e.g. tarsomere, a tarsal segment
metathorax	the last of the three thoracic segments and the one on which the wings are articulated
metepisternum	the side of the metathorax (*q.v.*)
monophagous	feeding on a single plant species
mucro	a terminal hook, tooth or spur on the inner side of the tibia
nodosity	a small swelling, bump or blunt process

nominate	name-bearing
non	not
oligophagous	feeding on few plants, usually those in a single genus
Orthoceri	a division of Curculionoidea, originally species with straight antennae
orthocerous	having straight, not geniculate (*q.v.*) antennae
oviposition	the act of egg-laying
ovoid	egg-shaped
papilla (pl. papillae)	a small nipple-like protuberance
papillate	bearing papillae (*q.v.*)
parthenogenesis	reproduction by females without fertilisation by males, usually because the latter are rare or non-existent
partim	in part
phanerognathous	with apparent mandibles, used of weevils with long rostra (*q.v.*) (opp. adelognathous)
phylogenetic	pertaining to evolutionary history
plastron	a permanent air-store in some aquatic weevils, maintained by a pile, or mat, of hydrofuge (*q.v.*) setae (*q.v.*)
polyphagous	feeding on many species of plant
polyphyletic	arising from several evolutionary lineages
porrect	extended forward
pronotum	the dorsal part of the prothorax (*q.v.*)
prosternum	the part of the prothorax (*q.v.*) visible from beneath
prothorax	the first thoracic segment
proximal	near (opp. distal)
pruina	a powdery deposit on the integument of certain weevils
pseudotetramerous	(of tarsi); apparently with four segments, the fifth being very small
pubescence	a pile, or covering, of fine setae (*q.v.*)
pygidium	the last dorsal abdominal segment
quadrate	as long as broad, square
recumbent	lying down, flat
remote	widely-spaced (of punctures, etc.) opp. to 'close'
resource partitioning	differential use of a resource (e.g. food-plant) by more than one species
reticulate	with a net-like pattern or structure
rostral	pertaining to the rostrum (*q.v.*)
rostrum (pl. rostra)	the extension of the head which bears the mouth-parts in weevils and is characteristic of them; also called 'beak' and 'snout'
rugose	rough, covered with small irregular ridges
s.l.	sensu lato; in the broad sense
s.str.	sensu stricto; in the strict sense
scale	a seta (*q.v.*) modified to form a small, thin, flat plate-like structure
scape	the first antennal segment
sclerite	a sclerotised (hardened) plate forming a segment or part of one
scrobe	the groove on the side of the rostrum (*q.v.*) which (usually) receives the scape (*q.v.*)

sensu	in the sense of
seta (pl. setae)	the 'hairs' or 'bristles' of weevils and insects generally
setose	bearing copious setae (*q.v.*)
shagreened	(of integument) minutely roughened, with fine sculpture
simple	unarmed, without appendages (e.g. fig. 212, p.81)
sinuate	winding, wavy-edged
stenophagous	feeding on a narrow range of plant species
sternite	a sclerotised plate of a sternum (*q.v.*)
sternum	the ventral part of a body segment
stria (pl. striae)	a longitudinal line on the elytra (*q.v.*) formed by a row of punctures or pits which may be joined or separate
subapical	immediately before the apex
substrate	that on which a species lives
subtruncate	nearly truncate (*q.v.*), not rounded or pointed
sulcus (pl. sulci)	a linear cavity or depression
supraorbital	above the (orbits of) the eyes
suture	the line between any two sclerites (*q.v*); specifically used of the line between the elytra (when at rest)
synanthropic	living in association with Man
taxon (pl. taxa)	a taxonomic entity the rank of which is not specified or stated
terete	smooth and cylindrical
tessellated, tessellation	with alternating pale and dark patches, mottled
transverse	broader than long
truncate	squared off
tubercle	a small, round, raised process, often pointed
tuberculate	covered with tubercles (*q.v.*)
unarmed	not bearing a tooth, spur or sharp process
uncus	the extension of the outer edge of the tibia to form a curved process, hook or tooth
unicolorous	of one colour
vagile	wandering, mobile
ventrite	a visible sternite (*q.v.*); in weevils the first ventrite is sternite 3, the first two being concealed
vertex	the dorsal surface of the head between the eyes
vestiture	clothing of pubescence (*q.v.*), scales (*q.v.*) or setae (*q.v.*) or any mixture of the three.

Curculionidae-Phanerognatha. Key to subfamilies

(Orthocerous groups, treated here as subfamilies but as families by Thompson (1992) and other workers, are indicated by an asterisk *. Names of groups not covered in the current handbook are in square brackets [ ]).

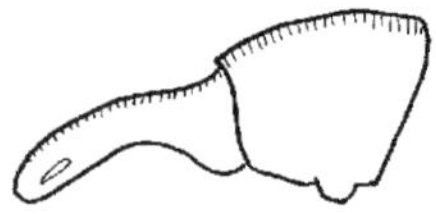

1   Weevils with eyes; tarsi 5-segmented (but pseudotetramerous in some cases); species living above ground.. **2**

-   Weevils without eyes (fig. 1); tarsi 4-segmented (fig. 2); living below ground [small species, 1.9-3.1 mm].
    .... **Raymondionyminae** *

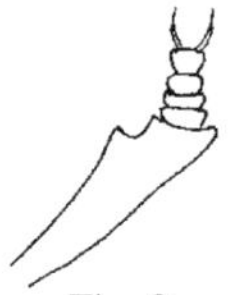

Fig. 1
head in lateral view

Fig. 2
fore-tarsus

2   Tarsi with two claws (exceptionally only one). **3**

-   Tarsi without claws (fig. 3) [small species, 1.7-2.8 mm].
    .... **Anoplinae**

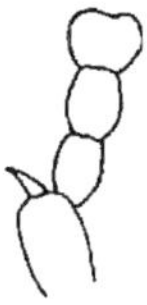

Fig. 3 tarsus

3   Club of antenna with three segments, all pubescent (fig. 4); base of scape fitting into scrobe when antenna retracted; [funiculus often with 7 segments]. **4**

-   Club of antenna apparently with two segments, the apical one small and pubescent, remainder of club glabrous and shining, not segmented (fig. 5); base of scape not fitting into scrobe when antenna retracted; [funiculus with either 4 or 6 segments]. .... **Dryophthorinae** *

Fig. 4
curculionine
antennal club

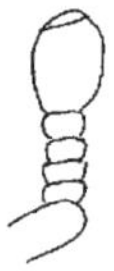

Fig. 5
dryophthorine
antennal club

4 Outer edge of fore-tibia prolonged smoothly and uninterruptedly into a hook, tooth, or long curved process ('uncus') (fig. 6) (to be distinguished carefully from any tooth ('mucro') on inner edge or middle of tibia, fig. 7). **5**

- Outer edge of fore-tibia not prolonged into a tooth or curved process.. **11**

5 Claw joint of tarsus very short, not, or scarcely, longer than preceding segment (fig. 8); [small species, 1.4-1.8 mm, associated with floating water plants]................ **Tanysphyrinae**

- Claw-joint of tarsus long, longer (usually much longer) than preceding segment (fig. 9).. **6**

6 Rostrum lying in a ventral thoracic groove, not directed forwards and extended into such a position only with difficulty.. **7**

- Rostrum porrect, evidently pointing forwards, easily seen from above, and not lying in a ventral thoracic groove........... **8**

7 Eyes close together on dorsum of head (fig. 10); hind femora usually dilated; [without erect or semi-erect scales].
.... **[Rhamphinae]**

- Eyes not close together on dorsum of head (fig. 11); hind femora not dilated; [often with some erect or semi-erect scales].. **Cryptorhynchinae**

8 Upper surface of pronotum and elytra with variegated, broad, erect scales (fig. 12); [antennal insertion in apical half of rostrum]................................ **Molytinae-Acicnemidini**

- Upper surface with appressed scales only, or without scales... **9**

9 Upper surface with at most fine setae or a few small tufts of pubescence. **10**

- Upper surface with patches of recumbent, appressed scales, or whole surface scaled............... **Molytinae-Pissodini**

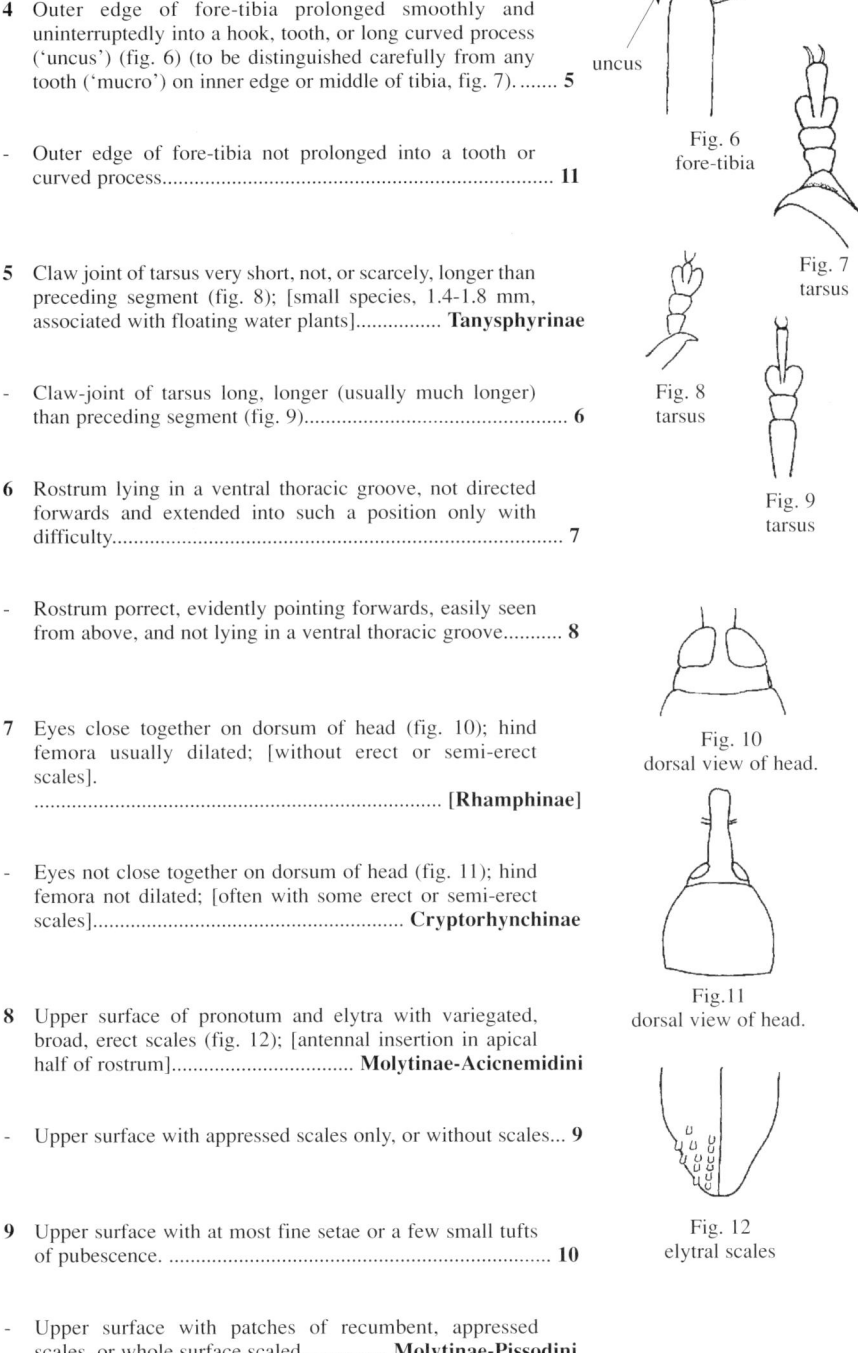

uncus

Fig. 6
fore-tibia

Fig. 7
tarsus

Fig. 8
tarsus

Fig. 9
tarsus

Fig. 10
dorsal view of head.

Fig.11
dorsal view of head.

Fig. 12
elytral scales

10 Elytra at base truncate or very weakly sinuate (fig. 13); base of scutellum, if present, level with base of elytra; fore-coxae widely separated (fig. 14); pronotum rounded at base, without lateral tubercles or swellings. .... **Cossoninae**

\- Elytral base strongly sinuate (fig. 15); base of scutellum lying behind extreme basal point of elytra; fore-coxae approximated (fig. 16); sides of pronotum sinuate, or divergent, at base, usually with lateral tubercles or swellings.. **Magdalidinae**

11 Rostrum less than twice (1.3-1.7 x) as long as broad (phanerognathous weevils with uncharacteristically short rostra: figs 17, 18).. **12**

\- Rostrum evidently elongate, more than twice as long as broad.. **13**

12 Small species, 1.9-2.1 mm; apex of rostrum red; [one aquatic species associated with water ferns, *Azolla* spp.] (fig. 17)....................................... **Erirhininae-Stenopelmini** *

\- Larger species, 5.0-7.0 mm; apex of rostrum black; [one terrestrial species associated with 'thistles', Asteraceae-Cardueae] (fig. 18). .... **Lixinae-Rhinocyllini**

13 Tarsal claws connate (fig. 19). **14**

\- Tarsal claws free (fig. 20). **17**

Fig. 13
base of pronotum and elytra

Fig. 14
prosternum

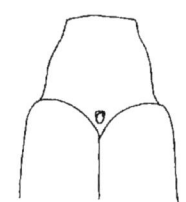

Fig. 15
base of elytra

Fig. 16
prosternum

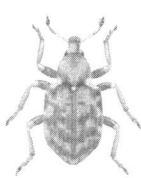

Fig. 17

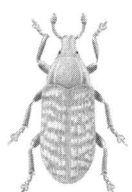

Fig. 18

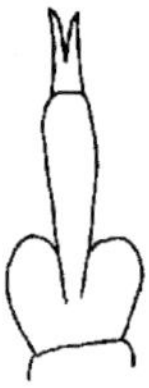

Fig. 19
connate tarsal claws

Fig. 20
free tarsal claws

14 Antennal funiculus with 6 or 7 segments. **15**

- Antennal funiculus with 5 segments (cf. fig. 21). **16**

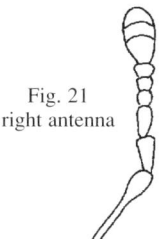

Fig. 21
right antenna

15 Small species, not longer than 2.5 mm; rostrum longer than pronotum; [associated with *Cuscuta* and *Centaurium* spp.]... **Smicronychinae**

- Larger species, at least 4.0 mm long, often longer (to 15.0 mm); rostrum shorter than pronotum; [mostly associated with Asteraceae-Cardueae, Apiaceae or Chenopodiaceae]. .... **Lixinae - Cleonini** and **Lixini**

16 Less elongate species, elytra 1.1-1.3 x as long as broad, with discrete black markings; abdominal segments 2-4 strongly curved posteriad at sides (fig. 22) [on average larger, 2.8-5.0 mm]. .... **Cioninae**

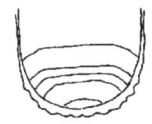

Fig. 22
abdominal segments.

- More elongate species, elytra 1.4-2.3 x as long as broad, generally unicolorous; abdominal segments straight or regularly curved at apex (fig. 23) [on average smaller, 1.3-4.0 mm]. [**Gymnetrinae** (*Mecinus* and *Gymnetro*n)]

17 Antennal funiculus with 5 segments (fig. 21: cf. couplet 14). .... [**Gymnetrinae** (*Miarus*)]

- Antennal funiculus with 6 or 7 segments............................ **18**

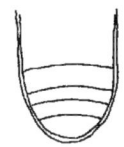

Fig. 23
abdominal segments

18 Disc of elytra with conspicuous ridges and/or nodosities; pronotum deeply pitted or tuberculate. **19**

- Disc of elytra without ridges; pronotum without deep pits or tubercles. **20**

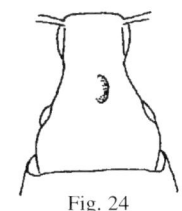

Fig. 24
head

19 Elytra with alternate interstices raised into conspicuous ridges, sides divergent to base, humeri strongly marked; pronotum with a deep median longitudinal depression; head deeply excavate between eyes and heavily ridged over them (fig. 24); smaller, 3.1-4.7 mm. .... **Cyclominae**

- Elytra with large irregular nodosities, sides parallel, sinuate at base (fig. 25), humeri wanting or very weakly developed; head with a weak depression between eyes, without ridges over them; on average larger (but variable) 4.5-7.2 mm... **Molytinae-Phrynixini**

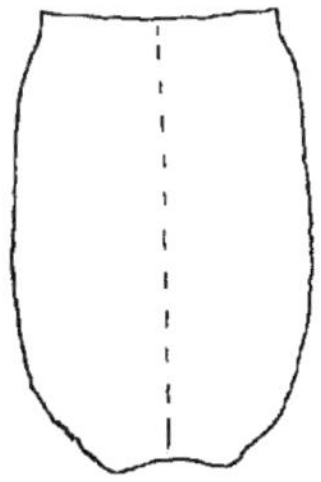

Fig. 25
outline of elytra

20 Third tarsal segment deeply divided, strongly bilobed (fig. 26) or, if not, then tarsi with claws as long as, or very little shorter than, tibiae.. **21**

- Third tarsal segment not deeply divided, not, or very weakly bilobed (fig. 27), and tarsi with claws much shorter than tibiae.. **Bagoinae**

21 Rostrum dilated in apical half; scrobes visible from above (fig. 28)... **Molytinae** (in part)

- Rostrum not dilated in apical half; scrobes not visible from above. **22**

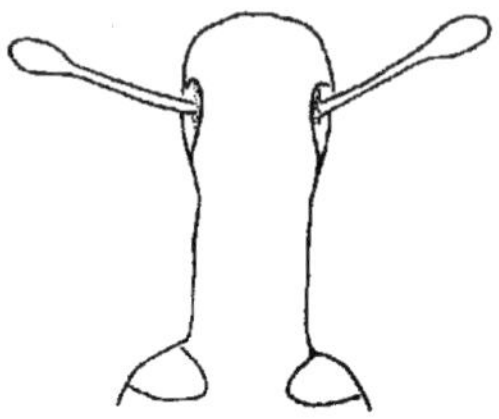

22 Fore-tibia without a mucro (internal tooth, or hook, at apex) (fig. 29) or, if with one, then pronotum with a conspicuous, raised, sharp anterior border. **23**

- Fore-tibia with an evident mucro (though sometimes obscured by setae) (fig. 26); pronotum without a conspicuous raised anterior border. **24**

23 Metathoracic epimera invisible from above (flat and hidden beneath elytral humeri); elytra more depressed and more elongate, 1.35-1.5 x as long as broad, and more nearly parallel-sided; [one species, small, 1.9-2.5 mm; legs mostly red, clothed with fine golden pubescence].
.... **[Tychiinae - Acalyptini]**

- Metathoracic epimera visible from above (fig. 31), raised and upstanding dorsad, not hidden by elytral humeri; elytra not, or less, depressed, generally less elongate, 1.0-1.3 x as long as broad, rounded at sides (a very few Ceutorhynchinae up to 1.5 x as long as broad).................... **25**

24 Tarsal claws toothed (fig. 30) or appendiculate (fig. 32)...... **26**

- Tarsal claws simple (fig. 33). **28**

25 Elytra markedly convex, cordiform; pronotum very strongly narrowed to apex (fig. 34), sides regularly curved, without a raised anterior border; mid- and hind-coxae very widely separated; first ventrite short and divided into three parts by hind coxae [one species, 2.0-2.8 mm]. .. **[Orobitinae]**

- Elytra less strongly convex; pronotum less strongly narrowed to apex, or sides sinuate, or with a conspicuous raised apical border; mid- and hind-coxae less widely separated; first ventrite not divided by hind-coxae.
.... **[Ceutorhynchinae]**

Fig. 26
tarsus

Fig. 27
tarsus

Fig. 28
rostrum

Fig. 29
fore-leg

Fig. 30
tarsal claws

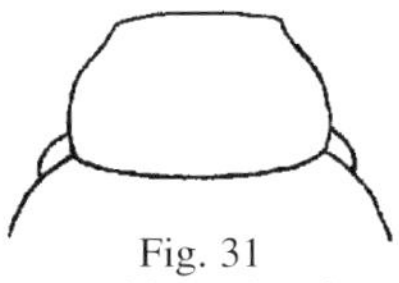

Fig. 31
elytral base showing humeri

Fig. 32
tarsal claws

Fig. 33
tarsal claws

26 Rostrum very long and slender (fig. 35); mandibles slender and pointed, moving vertically; tarsal claws toothed (fig. 30); antennae very long, distal funicular segments elongate (about twice as long as broad)...................... [**Curculioninae**]

- Rostrum shorter and more robust; mandibles more robust, moving horizontally; tarsal claws appendiculate (fig. 32); antennae shorter, distal funicular segments quadrate to transverse... **27**

Fig. 34
pronotum

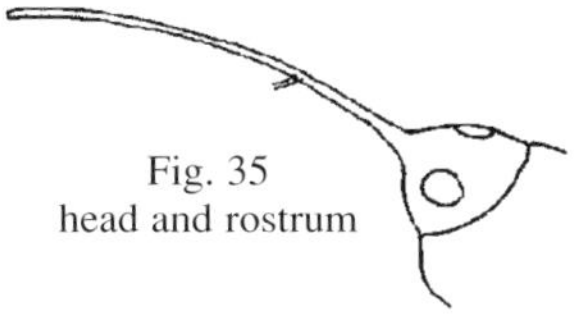

27 Eyes protuberant, round, asymmetrically domed or disc-like [species often arboreal]. .... [**Anthonominae**]

- Eyes normal, flat to moderately protuberant [species mainly associated with herbs]. .... [**Tychiinae**]

Fig. 35
head and rostrum

28 Metathoracic epimera not visible from above, obscured by pronotal base or elytral humeri. **29**

- Metathoracic epimera visible from above, not completely obscured by pronotal base or elytral humeri; [elongate species, 2.3-4.5 mm, predominantly shining black or bluish-black].. [**Baridinae**]

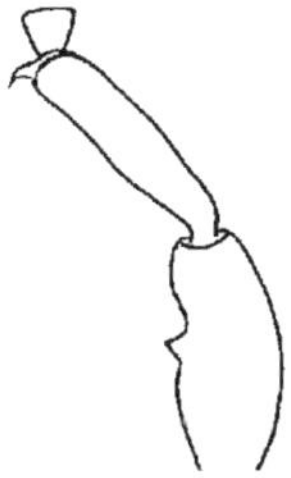

Fig. 36
fore-leg

29 Fore-femora with a broad median tooth (fig. 36) [arboreal, on *Salix* and *Populus* spp.].................................. **Dorytominae**

- Fore-femora unarmed (fig. 37).. **30**

30 Elytra with alternate interstices raised, each bearing a row of broad, conspicuous, erect setae........................... **Styphlinae**

- Elytra without raised interstices or erect setae..................... **31**

Fig. 37
fore-leg

31 Pronotum strongly transverse, as broad as elytra at shoulders (fig. 38); 10th elytral stria evanescent [very rare species, southern England only, on *Lotus*]............... **Storeinae**

- Pronotum less strongly transverse, narrower than elytra at shoulders; 10th elytral stria entire [including common and widely-distributed spp.]................... **Erirhininae-Erirhinini** *

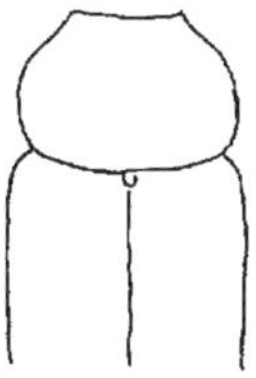

Fig. 38
pronotum and
base of elytra

Checklist

* Species of doubtful status, mainly unestablished introductions or possible migrants
† Extinct species

CURCULIONOIDEA Latreille, 1802

CURCULIONIDAE Latreille, 1802

RAYMONDIONYMINAE Reitter, 1912

FERRERIA Alonso-Zarazaga & Lyal, 1999
  RAYMONDIONYMUS auctt., non Wollaston, 1873
marqueti (Aubé, 1863)

DRYOPHTHORINAE Schoenherr, 1825

DRYOPHTHORINI Schoenherr, 1825

DRYOPHTHORUS Germar, 1824
corticalis (Paykull, 1792)

LITOSOMINI Lacordaire, 1866

SITOPHILUS Schoenherr, 1838 [ICZN 1959]
  CALANDRA Gistel, 1848, non [Clairville],1798
granarius (Linnaeus, 1758)
oryzae (Linnaeus, 1763)
zeamais Motschulsky, 1855

ERIRHININAE Schoenherr, 1825

ERIRHININI Schoenherr, 1825

ERIRHINUS Schoenherr, 1825
aethiops (Fabricius, 1792)

GRYPUS Germar, 1817
  GRYPIDIUS Schoenherr, 1826
equiseti (Fabricius, 1775)

NOTARIS Germar, 1817
acridulus (Linnaeus, 1758)
bimaculatus (Fabricius, 1787)
scirpi (Fabricius, 1792)

PROCAS Stephens, 1831
armillatus (Fabricius, 1801)
granulicollis J. Walton, 1848

THRYOGENES Bedel, 1884
festucae (Herbst, 1795)
fiorii Zumpt, 1928
nereis (Paykull, 1800)
schirrhosus (Gyllenhal, 1836)

STENOPELMINI Le Conte, 1876

STENOPELMUS Schoenherr, 1835
rufinasus Gyllenhal, 1835

LIXINAE Schoenherr, 1823
  CLEONINAE Schoenherr, 1826

CLEONINI Schoenherr, 1826

BOTHYNODERES Schoenherr, 1823
  CHROMODERUS Motschulsky, 1860
*affinis (Schrank, 1781)

CLEONIS Dejean, 1821
  CLEONUS Schoenherr, 1826
pigra (Scopoli, 1763)
  sulcirostris (Linnaeus, 1767)

CONIOCLEONUS Motschulsky, 1860
†hollbergi (Fahraeus, 1842)
nebulosus (Linnaeus, 1767)

LIXINI Schoenherr, 1823

LARINUS Germar, 1824
planus (Fabricius, 1792)

LIXUS Fabricius, 1801
?†angustatus (Fabricius, 1775)
  algirus auctt., non (Linnaeus, 1758)
†elongatus (Goeze, 1777)
†iridis Olivier, 1807
?†paraplecticus (Linnaeus, 1758)
scabricollis Boheman, 1843
?†vilis (Rossi, 1790)

RHINOCYLLINI Lacordaire, 1863

RHINOCYLLUS Germar, 1819
conicus (Frölich, 1792)

HYPERINAE Lacordaire, 1863

HYPERA Germar, 1817
  PHYTONOMUS Schoenherr, 1823
arator (Linnaeus, 1758)
†arundinis (Paykull, 1792)
dauci (Olivier, 1807)
diversipunctata (Schrank, 1798)
fuscocinerea (Marsham, 1802)
meles (Fabricius, 1792)
nigrirostris (Fabricius, 1775)
ononidis Chevrolat, 1863
pastinacae (Rossi, 1790)

26

plantaginis (De Geer, 1775)
pollux (Fabricius, 1801)
   adspersa (Fabricius, 1792) non (Fabricius, 1775)
postica (Gyllenhal, 1813)
punctata (Fabricius, 1775)
   ?*austriaca* (Schrank, 1781)
   ?*zoilus* (Scopoli, 1763)
rumicis (Linnaeus, 1758)
suspiciosa (Herbst, 1795)
venusta (Fabricius, 1781)

LIMOBIUS Schoenherr, 1843
borealis (Paykull, 1792)
mixtus (Boheman, 1834)

CIONINAE Schoenherr, 1825

CIONUS Clairville, 1798
alauda (Herbst, 1784)
hortulanus (Fourcroy, 1785)
longicollis C. Brisout, 1863
nigritarsis Reitter, 1904
scrophulariae (Linnaeus, 1758)
tuberculosus (Scopoli, 1763)

CLEOPUS Dejean, 1821
pulchellus (Herbst, 1795)

MOLYTINAE Schoenherr, 1823
   HYLOBIINAE W. Kirby, 1837

MOLYTINI Schoenherr, 1823
   LIPARINI Latreille, 1828

MOLYTINA Schoenherr, 1823

LIPARUS Olivier, 1807
coronatus (Goeze, 1777)
germanus (Linnaeus, 1758)

LEIOSOMATINA Reitter, 1913

LEIOSOMA Stephens, 1829
deflexum (Panzer, 1795)
oblongulum Boheman, 1842
troglodytes Rye, 1873
   pyrenaeum sensu auctt. partim, non C. Brisout, 1866

PLINTHINA Lacordaire, 1863

MITOPLINTHUS Reitter, 1897
   EPIPOLAEUS Weise, 1907
   PLINTHUS sensu auctt. partim, non Germar, 1817
caliginosus (Fabricius, 1775)

TYPODERINA Voss, 1965

ANCHONIDIUM Bedel, 1884
unguiculare (Aubé, 1850)

ACICNEMIDINI Lacordaire, 1866
   TRACHODINI Le Conte, 1876

TRACHODES Germar, 1824
hispidus (Linnaeus, 1758)

HYLOBIINI W. Kirby, 1837

HYLOBIUS Germar, 1817
abietis (Linnaeus, 1758)
transversovittatus (Goeze, 1777)

LEPYRINI W. Kirby, 1837

LEPYRUS Germar, 1817
†*capucinus* (Schaller, 1783)

PHRYNIXINI Kuschel, 1964

SYAGRIUS Pascoe, 1875
intrudens C. O. Waterhouse, 1903

PISSODINI Gistel, 1848

PISSODES Germar, 1817
castaneus (De Geer, 1775)
plni (Linnaeus, 1758)
validirostris (C. R. Sahlberg, 1834)

CYCLOMINAE Schoenherr, 1826

RHYTHIRRININI Lacordaire, 1863

GRONOPINA Bedel, 1884

GRONOPS Schoenherr, 1823
inaequalis Boheman, 1842
lunatus (Fabricius, 1775)

MESOPTILIINAE Lacordaire, 1863
   MAGDALIDINAE auctt.

MAGDALIDINI Pascoe, 1870

MAGDALIS Germar, 1817

Sg. *ODONTOMAGDALIS* Barrios, 1984
   MAGDALINUS sensu auctt., non Germar, 1843
armigera (Fourcroy, 1785)
carbonaria (Linnaeus, 1758)

Sg. *MAGDALIS* s.str.
   MAGDALINUS Germar, 1843
duplicata Germar, 1818
memnonia (Gyllenhal in Faldermann, 1837)
phlegmatica (Herbst, 1797)

Sg. *EDO* Germar, 1819
ruficornis (Linnaeus, 1758)

27

Sg. *PORROTHUS* Dejean, 1821
  NEOPANUS Reitter, 1916
cerasi (Linnaeus, 1758)

Sg. *PANUS* Schoenherr, 1826
barbicornis (Latreille, 1804)

ANOPLINAE Bedel, 1884

ANOPLUS Germar, 1820
plantaris (Naezen, 1794)
roboris Suffrian, 1840

COSSONINAE Schoenherr, 1825

COSSONINI Schoenherr, 1825

COSSONUS Clairville, 1798
linearis (Fabricius, 1775)
parallelepipedus (Herbst, 1795)

RHOPALOMESITES Wollaston, 1873
  MESITES auctt. Brit., non Schoenherr, 1838
tardyi (Curtis, 1825)

DRYOTRIBINI Le Conte, 1876

CAULOPHILUS Wollaston, 1854
*oryzae (Gyllenhal, 1838)

ONYCHOLIPINI Wollaston, 1873

PSELACTUS Broun, 1886
spadix (Herbst, 1795)

PSEUDOPHLOEOPHAGUS Wollaston, 1873
  CAULOTRUPODES Voss, 1955
aeneopiceus (Boheman, 1845)

STEREOCORYNES Wollaston, 1873
truncorum (Germar, 1824)

PENTARTHRINI Lacordaire, 1866

EUOPHRYUM Broun, 1909
confine (Broun, 1881)
rufum (Broun, 1880)

PENTARTHRUM Wollaston, 1854
huttoni Wollaston, 1854

RHYNCOLINI Gistel, 1848

RHYNCOLINA Gistel, 1848

MACRORHYNCOLUS Wollaston, 1873
*littoralis (Broun, 1880)

RHYNCOLUS Germar, 1817 [ICZN 1991b]
  EREMOTES Wollaston, 1861
ater (Linnaeus, 1758)

PHLOEOPHAGINA Voss, 1955

PHLOEOPHAGUS Schoenherr, 1838
†gracilis (Rosenhauer, 1856)
lignarius (Marsham, 1802)

CRYPTORHYNCHINAE Schoenherr, 1825

CRYPTORHYNCHINI Schoenherr, 1825

CRYPTORHYNCHINA Schoenherr, 1825

CRYPTORHYNCHUS Illiger, 1807
lapathi (Linnaeus, 1758)

TYLODINA Lacordaire, 1866

ACALLES Schoenherr, 1825
misellus Boheman, 1844
  turbatus sensu auctt., non Boheman, 1844
ptinoides (Marsham, 1802)
roboris Curtis, 1835

TANYSPHYRINAE Gistel 1848

TANYSPHYRUS Germar, 1817
lemnae (Paykull, 1792)

BAGOINAE C. G. Thomson, 1859

BAGOUS Germar, 1817

Sg. *HYDRONOMUS* Schoenherr, 1825
alismatis (Marsham, 1802)

Sg. *EPHIMEROPUS* Hochhuth, 1847
†petro (Herbst, 1795)

Sg. *CYPRUS* Schoenherr, 1825
tubulus Caldara & O'Brien, 1994
  angustus Silfverberg, 1977, non Tanner, 1954
  cylindrus (Paykull, 1800), non (Fabricius, 1781)

Sg. *BAGOUS* s. str.
argillaceus Gyllenhal, 1836
†binodulus (Herbst, 1795)
brevis Gyllenhal, 1836
collignensis (Herbst, 1797)
  claudicans auctt. Brit., non Boheman, 1845
czwalinae Seidlitz, 1891
?†diglyptus Boheman, 1845
frit (Herbst, 1795)
limosus (Gyllenhal, 1827)
longitarsis C.G. Thomson, 1868
  arduus Sharp, 1917

lutulosus (Gyllenhal, 1827)
nodulosus Gyllenhal, 1836
subcarinatus Gyllenhal, 1836
tempestivus (Herbst, 1795)

Sg. *ABAGOUS* Sharp, 1916
glabrirostris (Herbst, 1795)
lutosus (Gyllenhal, 1813)
lutulentus (Gyllenhal, 1813)
puncticollis Boheman, 1845
?†*robustus* H. Brisout, 1863
    rudis Sharp, 1917

DORYTOMINAE Bedel, 1886

DORYTOMINI Bedel, 1886

DORYTOMUS Germar, 1817
affinis (Paykull, 1800) non (Schrank, 1781)
    ?*edoughensis* Desbrochers, 1875
dejeani Faust, 1882
filirostris (Gyllenhal, 1836)
hirtipennis Bedel, 1884
ictor (Herbst, 1795)
    validirostris (Gyllenhal, 1836)
longimanus (Forster, 1771)
majalis (Paykull, 1800)
melanophthalmus (Paykull, 1792)
rufatus (Bedel, 1888)
salicinus (Gyllenhal, 1827)
salicis J. Walton, 1851

taeniatus (Fabricius, 1781)
tortrix (Linnaeus, 1761)
tremulae (Fabricius, 1787)

STOREINAE Lacordaire, 1863

STOREINI Lacordaire, 1863

PACHYTYCHIUS Jekel, 1861
haematocephalus (Gyllenhal, 1836)

STYPHLINAE Jekel, 1861

STYPHLINI Jekel, 1861

ORTHOCHAETES Germar, 1824
insignis (Aubé, 1863)
setiger (Beck, 1817)

PSEUDOSTYPHLUS Tournier, 1874
pillumus (Gyllenhal, 1836)

SMICRONYCHINAE Seidlitz, 1891

SMICRONYCHINI Seidlitz, 1891

SMICRONYX Schoenherr, 1843
coecus (Reich, 1797)
jungermanniae (Reich, 1797)
reichi (Gyllenhal, 1836)
    seriepilosus Tournier, 1874

Subfamily Raymondionyminae

Blind, hypogean (soil-inhabiting) weevils are known in eight curculionoid subfamilies (Osella, 1979), but only the Raymondionyminae consist entirely of such species. As indicated earlier, the group is currently regarded as a family of the Curculionoidea-Orthoceri (Thompson, 1992), but is treated here as a curculionid subfamily in order not to disturb unduly the classificatory arrangement of Pope (1977). Fifteen genera are currently included in the Raymondionyminae and the group has a predominently Mediterranean distribution, though this may in part reflect the area in which most work on these obscure and rare weevils has been concentrated. Representatives are known also from other parts of the southern Palaearctic region, California, Madagascar and New Zealand. The group was monographed by Osella (1977).

The capture of a specimen of *Raymondionymus marqueti* in Kew Gardens (Williams, 1968) was, from the location, thought to be either a chance introduction or a representative of an established species. However, specimens were taken sparingly, first in excavated roots and then in deep pitfall traps, at Bromley, Kent, by Thompson (1995). Since then it has been taken commonly at Epsom and neighbouring areas of Surrey, in deep pitfall traps of a particular design (Owen, 1995; 1997) and is either a native British species or, more probably, a long-established, possibly synanthropic, alien.

It has recently been established that the inclusion of our species in *Raymondionymus* was based on a misidentification (Alonso-Zarazaga & Lyal, 1999), and it is now placed in *Ferreria*.

Genus *Ferreria* Alonso-Zarazaga & Lyal 1999

F. marqueti exists as five subspecies or allopatric forms, one of questionable validity (Osella, 1977). British specimens are ssp. *marqueti s. str.* (G. Osella, *pers. comm.*; Owen, 1997). This nominate subspecies occurs in France, including the north.

- One species; a small (1.9-3.1 mm), hypogean, brown, eyeless (fig. 39), deeply punctured weevil with conspicuously flattened and expanded tibiae (fig. 40). .... *marqueti*

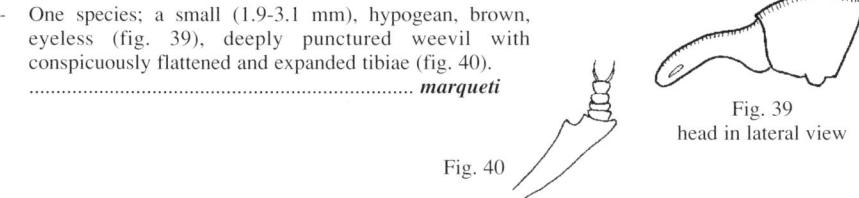

Fig. 39
head in lateral view

Fig. 40

Male with a deep ventral abdominal fossa or depression. Female without, or with a much shallower fossa.

A hypogean species of obscure and little-known habits. It probably feeds on dead vegetable matter and living roots and may be polyphagous (like most root-feeding weevils). However, it seems to be associated mainly with introduced ornamental conifers. Rare and very local, though taken in considerable numbers at several sites in Surrey and W Kent, also recorded from Dorset and Middlesex (Highgate). Only recently discovered as an established, breeding species and likely to be taken only, or mainly, in specialised and deeply set pitfall traps (Kuschel, 1991; Thompson, 1995; Owen, 1995; 1997). The species is known from France, Switzerland, Italy, Corsica, Slovenia, Croatia, Bosnia, Herzegovina and Tunisia. Its discovery as a breeding species in England represents a considerable northwards extension of its known range, and indeed that of Raymondionyminae as a whole.

Subfamily Dryophthorinae

As stated in the introduction, this group is currently regarded as a family of Curculionoidea-Orthoceri; the family-group name Rhynchophoridae (Morimoto, 1962a; 1962b; Thompson, 1992; Zimmerman, 1993) has been replaced by Dryophthoridae, which

30

is prior. In the British Isles two genera are represented, *Dryophthorus* and *Sitophilus*. Their higher classification in family Dryophthoridae (as Rhynchophoridae) is given by Zimmerman (1993). Here, for convenience, the genera are considered to be placed in tribes Dryophthorini and Litosomini respectively.

Several species of Dryophthorinae are important pests of tropical crops. One species, *Cosmopolites sordidus* (Germar) ('banana root weevil'), became established for a brief period in a York hothouse (Blair, 1948), but is not known to exist, even under artificial conditions, in the British Isles today.

Key to tribes and genera

Pygidium conspicuously exposed, with a dorsal sulcus (fig. 41); elytra simple at apex; funiculus with six segments (fig. 42); tarsi 4-segmented; pronotum very large, strongly elongate, broader than elytra and, together with head, as long, or nearly as long, as elytra; rostrum strongly dilated at antennal insertion [stored products pests].................................... **Litosomini**, one genus, *Sitophilus*

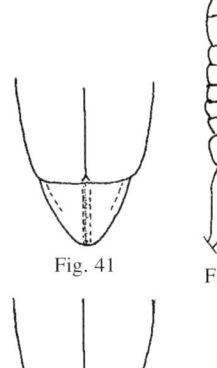

Fig. 41

Fig. 42

- Pygidium hidden by elytra; elytra with a conspicuous lateral expansion or flange at apex (fig. 43); funiculus with four segments (fig. 44); tarsi 5-segmented; pronotum not large, slightly elongate, narrower than elytra and, together with head, less than two-thirds as long as elytra; rostrum slightly dilated, but in front of antennal insertion [in dead trees in the open].

Dryophthorini, one genus, *Dryophthorus*

Fig. 43

Fig. 44

Genus *Sitophilus* Schoenherr

The three species of this genus established in the British Isles are cosmopolitan pests which probably all originated in the Oriental region. A graphic account of the depredations and abundance of *S. oryzae* is given in Zimmerman (1993). The British species seem to be less common than formerly because of better hygiene in the storage and preparation of farinaceous foods. There is an extensive literature on all aspects of the morphology, physiology and biology of the species and particularly on their control.

Key to species

1 Pronotal disc with close, isodiametric punctures (fig. 45); punctures of elytral striae broad and isodiametric, striae broader than interstices; metepisternum as broad as adjacent outermost elytral interstice [fully winged]............... **2**

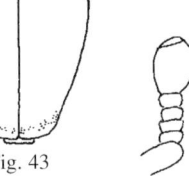

Fig. 45

- Pronotal disc with somewhat remote, elongate, oval punctures (fig. 46); punctures of elytral striae narrow and elongate, striae narrower than interstices; metepisternum narrow, narrower than adjacent outermost elytral interstice [flightless, with reduced wings; length 3.1-4.3 mm] .... *granarius*

Fig. 46

Male rostrum a little shorter, broader and less strongly curved, more closely and coarsely punctured, and a little duller than female.

In grain stores, formerly in shops, kitchens, stables etc. An important pest ('granary weevil') of grain, including wheat, barley, oats, maize and birdseed etc; also in farinaceous foodstuffs such as biscuits. Also in non-grain seeds for instance acorns, at least under experimental conditions. Larvae in individual grains (seeds). Continuously brooded. Local, and less common than formerly, but widely distributed throughout Enland northwards to W Perth. Few Welsh records. Very local in Ireland. Most records are from buildings in towns, but the species has also been taken outdoors, for example in association with grain fed to game birds. Almost completely cosmopolitan.

2   Length normally less than 3.0 mm; generally slightly redder; a little duller, microsculpture of pronotal and elytral interspaces coarser; median lobe without longitudinal dorsal sulci, more smoothly convex dorsally in cross-section (fig. 47) [length 2.3-3.9 mm]. .... ***oryzae***

Fig. 47

Male rostrum shorter and broader, rougher and less strongly shining than female.
In granaries etc. and generally in similar premises and products as those in which S. granarius *is found. An even more damaging pest ('rice weevil') than that species because of its ability to fly and infest new grain supplies. Larvae in individual grains. Continuously brooded. Following the discovery that the two 'strains' of rice weevil (Richards, 1944) are separate species (Floyd & Newsom, 1959) the exact distribution of* S. oryzae *in the British Isles is uncertain.* S. oryzae *of older literature has been recorded from much of England from the south coast northwards to Stirling, from Glamorgan and a few Irish vice-counties, but some of these records undoubtedly should refer to* S. zeamais. *The latter species is regarded as 'more feral' (less synanthropic) than* S. oryzae, *at least in Australia (Zimmerman, 1993), and examination of only a few old specimens in museum collections suggests that* S. zeamais *has been standing under the name of* S. oryzae *in several cases. Cosmopolitan (Halstead, 1964).*

-   Length normally more than 3.0 mm; generally darker and less red; slightly less dull, microsculpture of pronotal and elytral interspaces finer; median lobe with two dorsal longitudinal sulci, dorsal surface sinuate in cross-section (fig. 48) [length 2.6-4.2 mm]. .... ***zeamais***

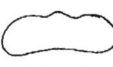

Fig. 48

Male rostrum shorter (shorter than width of pronotum), more robust, much more strongly punctured, dull, and usually with a median longitudinal keel. Female rostrum longer (about as long as pronotum is wide), more slender, less strongly punctured, shining, and without a median longitudinal keel.
In grain stores etc., and generally in similar premises to those inhabited by other Sitophilus *spp. 'More feral' status (Zimmerman, 1993) is not definitely established for the British Isles. A serious pest ('maize weevil') which, like* S. oryzae, *readily infests grain by flight. Larvae in individual grains. Continuously brooded. Status, abundance and distribution in the British Isles uncertain, but the few definite records (e.g. Huntingdon, Cardigan) do not suggest that it is a species occurring only under artificial conditions (Pope, 1977). It is probable that* S. oryzae *and* S. zeamais *are mixed in British collections and that records are consequently also mixed. Cosmopolitan (Halstead, 1964).*

Genus *Dryophthorus* Germar

Although long classified in Cossoninae (e.g. Pope, 1977), *Dryophthorus* is currently included with *Sitophilus* in Dryophthorinae (correctly Dryophthoridae, as stated in the Introduction). Its affinities are discussed by Zimmerman (1993). Representatives of the genus are found in all zoogeographical regions, except the Afrotropical, though some species may have been spread by commerce in timber. However, only one species, which occurs in Britain, is found in the Palaearctic.

-   One species; a small (3.2-3.6 mm) dark brown to black weevil; however specimens are often encrusted with debris; elytral interstices raised; pronotum constricted at about 1/4 from apex; other characters in generic key, p. 31 .... ***corticalis***

32

Male rostrum a little shorter and broader; antennal insertion in basal third of rostrum. Female rostrum slightly longer and narrower; antennal insertion in basal quarter of rostrum.

In ancient forest and pasture woodland (parkland). In dead, rotten wood of dead standing and fallen oaks. Sometimes in association with old tunnels of the ant Lasius brunneus *(Latreille). Also occasionally in or on dead wood of other tree species. Extremely local in the British Isles, though it can be abundant when found. Known only from Windsor Forest and Windsor Great Park, Berks. It was first found by Donisthorpe (1925) but lost sight of between 1936 and 1980, when it was rediscovered by Owen (1983) and subsequently found by other coleopterists. Throughout most of Europe, including the north, but generally scarce. RDB1, the species' survival threatened by 'tidying up' operations which destroy the dead wood which it inhabits (Owen, 1983).*

Subfamily Erirhininae

In this *Handbook* the 'orthocerous' family Erirhinidae (Thompson, 1992) is treated as a subfamily of Curculionidae, as stated in the Introduction. Its composition is reduced compared with the genera included in the subfamily by Pope (1977). The 'gonatocerous' groups removed to other subfamilies are *Bagous* (*s.l.*; Bagoinae), *Dorytomus* (Dorytominae), *Pachytychius* (Storeinae), *Pseudostyphlus* and *Orthochaetes* (Styphlinae) and *Smicronyx* (Smicronychinae). The subfamily Tanysphyrinae of Pope (1977), sited somewhat uneasily between Anoplinae and Cossoninae, is retained in that position; recent work (M. V. L. Barclay pers. comm.) has established that it is not erirhinine.

The subfamily, as treated here, includes two tribes in the British fauna, Stenopelmini and Erirhinini. As they have been keyed out in the subfamily key (pp. 22, 25), no separate key to tribes is provided here.

Erirhininae are predominantly associated with semi-aquatic habitats, with only species of *Procas* occurring in strictly terrestrial ones. *Grypus*, being associated with *Equisetum*, is most frequently found in wet places, while all the other species occur on either floating or emergent aquatic plants.

Tribe Stenopelmini

The introduced North American genus *Stenopelmus* is the only representative of this tribe in the British fauna.

Genus *Stenopelmus* Schoenherr

This genus is monotypic.

-   One species; a small (2.3-2.9 mm) weevil, superficially resembling a *Pelonomus* or *Rhinoncus* species (Ceutorhynchinae); rostrum short, with a red apex (habitus fig. 49 overleaf); colour figure in Donisthorpe (1931)) ....*rufinasus*

Male first abdominal ventrite in midline flat or slightly concave. Female first abdominal ventrite convex. The character given by Hoffmann (1958) to distinguish the sexes is erroneous; the apical tibial spur, or hook, is present in both sexes, not just the male.

In ponds, lakes, canals, ditches, dykes and other still water-bodies. On water fern, Azolla filiculoides, *an introduced but well-established species. Larvae feed on the leaves, especially those above the waterline; the larva and life history in North America were described by Richerson & Grigarick (1967). Neither adult nor larva is well-adapted to aquatic life, although adults can walk on the water surface (cf.* Tanysphyrus lemnae). *First recorded in Britain (from the Norfolk Broads) in 1921 (Janson, 1921), since when it has been found sporadically in southern England and South Wales from S Devon eastwards to E Kent and northwards to Cambridge and E Norfolk; it has recently been recorded from Cumberland (Read, 1999). It is often abundant where found. A native to the southern*

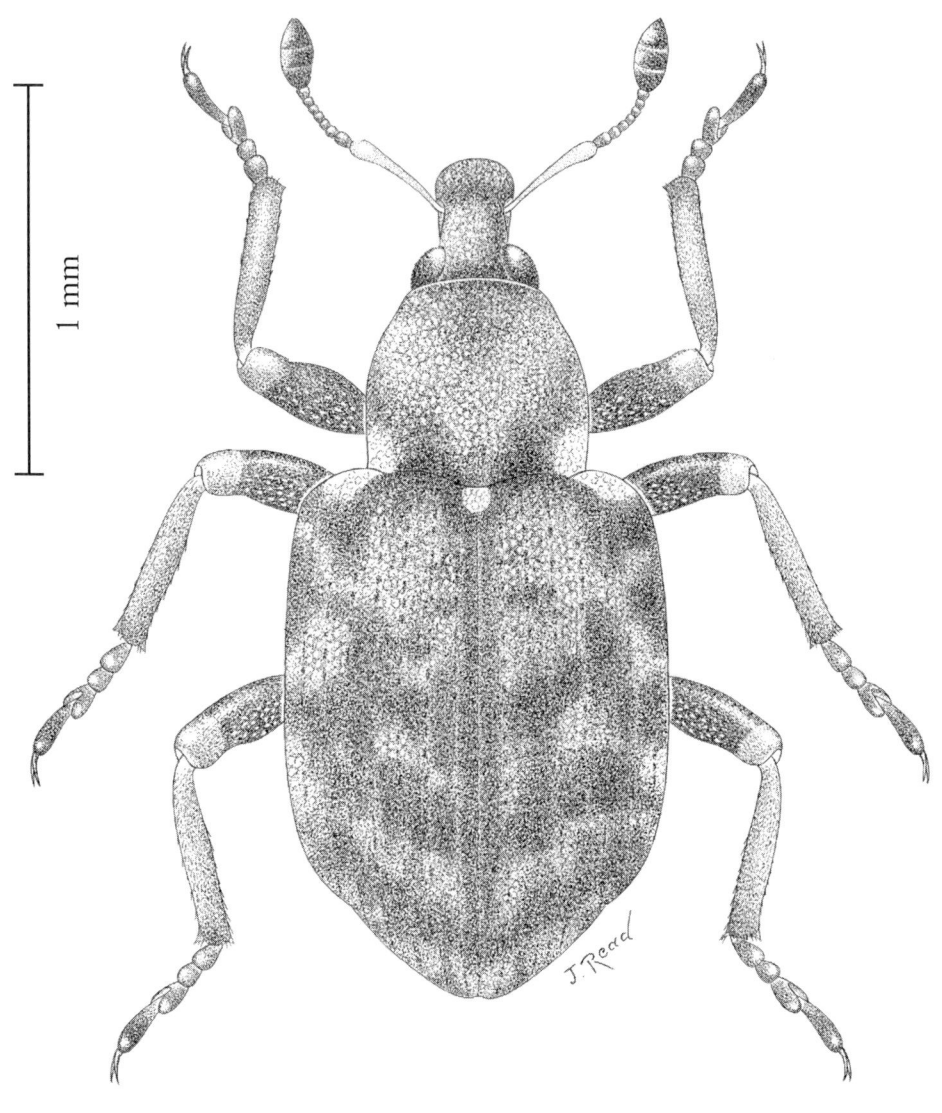

1 mm

Fig. 49
Stenopelmus rufinasus

U.S.A., it has been introduced into Europe with the host and found in France, Germany, the Low Countries and Italy, and also occurs in the Canary Islands (Gran Canaria: author, unpublished).

Tribe Erirhinini

This tribe includes most of the genera currently placed in Erirhininae (*s. l.*). The species are medium-sized (2.9-8.7 mm), mostly slightly depressed species, living in wetland habitats.

Key to genera

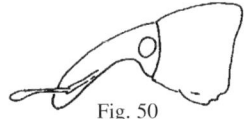

Fig. 50

1   Antennae inserted near apex of rostrum, at about 1/5 from apex (fig. 50), apical portion of rostrum (from antennal insertion) little longer than broad. .... *Procas*

-   Antennae inserted further from apex of rostrum, at about 1/4 or more from apex, apical portion of rostrum much longer than broad. **2**

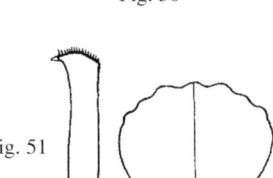

Fig. 51

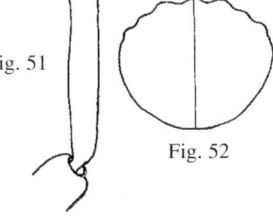

Fig. 52

2   Outer edge of fore-tibiae almost straight, not curved inwards at apex (fig. 51); elytra higher compared with breadth, about 1.1 x as broad as high (fig. 52), interstices 3, 5 and 7 interruptedly raised, apex and sides conspicuously paler than base and disc......................... *Grypus*

-   Fore-tibiae curved inwards at apex (figs 53, 54); elytra shallower, more depressed compared with breadth, about 1.25 x as broad as high, uneven-numbered interstices not interruptedly raised, though all interstices slightly convex in some species; elytra unicolorous or inconspicuously mottled, apex and sides not paler than base and disc. **3**

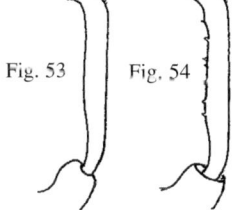

Fig. 53   Fig. 54

3   Interstice 1 of elytra at suture clothed with scales; anterior margin of prosternum not emarginate; elytra more elongate, 1.8-2.0 x as long as broad, more clearly parallel-sided; eyes more strongly rounded (fig. 55). ....*Thryogenes*

-   Interstice 1 of elytra bare, or at most with very sparse pubescence; anterior margin of prosternum emarginate; elytra less elongate, 1.4-1.65 x as long as broad, slightly rounded or less clearly sub-parallel at sides; eyes very weakly rounded, almost flat (fig. 56). **4**

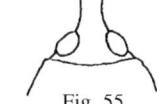

Fig. 55

Fig. 56

4   Pronotum closely to rugosely punctured, with a slightly raised smooth, shining, median longitudinal line (sometimes somewhat effaced); dorsum dull, usually with at least some distinct pubescence on pronotum and elytra [including species widespread in southern England]. ...*Notaris*

-   Pronotum with remote punctures (interspaces mostly wider than punctures themselves; fig. 57), without a smooth, shining median longitudinal line, or with a very obscure, inconspicuous and unraised one; dorsum strongly shining, glabrous [species of Northern England, Scotland and Ireland only].. *Erirhinus*

Fig. 57

Genus *Procas* Stephens

This is a small genus of about 16 European species most of which inhabit the Mediterranean region; there is also one North American species. They are mostly uncommon and little-known weevils of obscure habits. Two species inhabit the British Isles, including one of the few endemic British insects; they were discussed and compared in detail by Kenward (1990).

Key to species

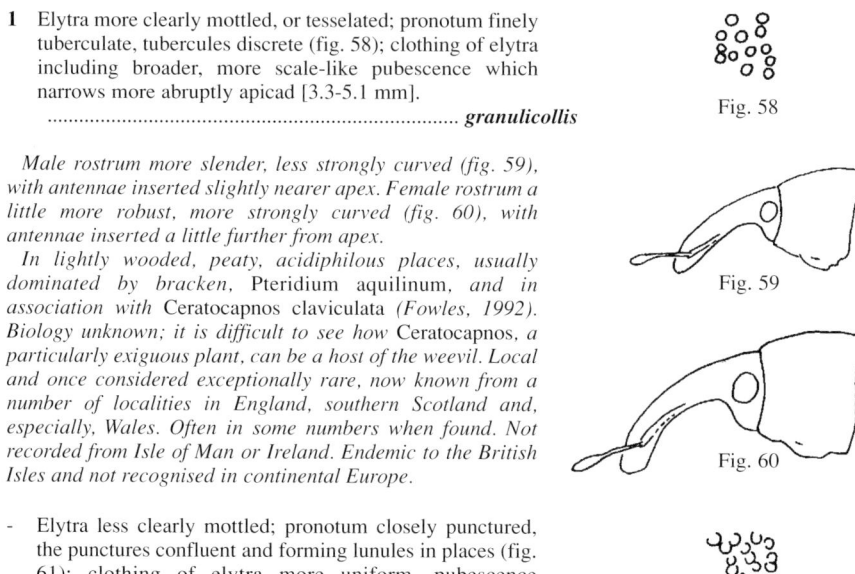

1   Elytra more clearly mottled, or tesselated; pronotum finely tuberculate, tubercules discrete (fig. 58); clothing of elytra including broader, more scale-like pubescence which narrows more abruptly apicad [3.3-5.1 mm].
.... *granulicollis*

Fig. 58

Male rostrum more slender, less strongly curved (fig. 59), with antennae inserted slightly nearer apex. Female rostrum a little more robust, more strongly curved (fig. 60), with antennae inserted a little further from apex.

In lightly wooded, peaty, acidiphilous places, usually dominated by bracken, Pteridium aquilinum, *and in association with* Ceratocapnos claviculata *(Fowles, 1992). Biology unknown; it is difficult to see how* Ceratocapnos, *a particularly exiguous plant, can be a host of the weevil. Local and once considered exceptionally rare, now known from a number of localities in England, southern Scotland and, especially, Wales. Often in some numbers when found. Not recorded from Isle of Man or Ireland. Endemic to the British Isles and not recognised in continental Europe.*

Fig. 59

Fig. 60

-   Elytra less clearly mottled; pronotum closely punctured, the punctures confluent and forming lunules in places (fig. 61); clothing of elytra more uniform, pubescence narrower, tapering regularly and finely to apex [3.8-6.8 mm]... *armillatus*

Fig. 61

Secondary sexual differences very slight. Male rostrum a little less regularly curved, curve somewhat abrupt from antennal insertion apicad; antennae inserted slightly closer to rostral apex. Female rostrum more smoothly curved, without an abrupt break in curvature from antennal insertion; antennae inserted slightly further from rostral apex.

In dry, open country, generally under stones and clods, but recorded from marshy places early in the 19th century and in such places abroad. The only British site where the weevil was taken abundantly was a field of oats near the sea, in the vicinity of Brighton, E Sussex. It was surmised that larvae were root-feeders, but no host was recorded (Cox, 1930); the biology of the species is not known either in Britain or elsewhere. Also recorded, generally as single individuals, from several vice-counties in southern England northwards to Nottingham. Western Europe and North Africa; Madeira, Canaries. RDB3.

N.B. Recent work on *Procas* by Mr R. T. Thompson will considerably modify this account.

Genus *Erirhinus* Schoenherr

Erirhinus and *Notaris* are treated here as separate genera, following Kuschel (unpublished), who has reviewed the world fauna. Accordingly, there is just a single British species in the genus.

- One species; a strongly shining, glabrous, black weevil, legs and antennae reddish [5.3-7.9 mm]......................*aethiops*

Sexual differences not marked. Male rostrum a little more strongly curved and with antennae inserted slightly nearer apex. Female rostrum slightly straighter, with antennae inserted a little further from apex.

In bogs, mires and wetlands generally, and at the margins of rivers and other watercourses; a glacial relict species. Traditionally associated with Sparganium erectum *in the British Isles, but also recorded from* Carex *spp. on some sites where* Sparganium *is absent. Biology unknown. Local, and usually regarded as rare, but quite widely distributed from NE Yorks. northwards to Easterness, though not recorded from northern Scotland nor any of the Islands. Widespread, though local, in Ireland. No records from Wales, southern England or Isle of Man. Eastern, northern and central Europe (montane in the south of the region); North America.*

Genus *Notaris* Germar

The recognition of separate generic assignment for *E. aethiops* leaves three British species in *Notaris*, a genus of about 20 Palaearctic species, with a further three in North America.

Key to species

1 Inner edge of fore-tibia smooth, without a trace of any tubercules or blunt teeth (fig. 62); elytra without a lineate pattern, pubescence, if present, not denser or longer on alternate interstices. **2**

- Inner edge of fore-tibia with 4-7 conspicuous tubercules or blunt teeth (fig. 63); elytra with a lineate pattern, pubescence denser and longer on alternate interstices (sometimes obscurely so) [5.5-8.7 mm].................*bimaculatus*

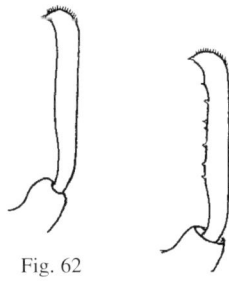

Fig. 62

Fig. 63

Male antennae inserted closer to rostral apex, apical portion of rostrum (from insertion) shorter than scape; elytra narrower and less rounded at sides. Female antennae inserted further from rostral apex (cf. fig. 64), apical portion of rostrum as long as, or longer than, scape; elytra broader and more rounded at sides.

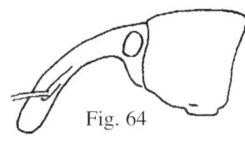

Fig. 64

In marshes, including brackish ones, and at the sides of rivers and other water courses. Apparently polyphagous on a wide range of emergent and semi-aquatic Monocotyledons (Poaceae (=Graminae), Cyperacae and Typhaceae) but biology and foodplants little studied in Britain. Putative British hosts include Phalaris arundinacea, Phragmites australis, Typha latifolia *and* Carex *spp. Adult weevils have been found in apparent larval feeding cavities in the shoots of* Glyceria maxima *and a* Carex *sp. in Germany. Local and rather rare, but widely distributed throughout England, Wales and southern Scotland northwards to Ayr and Berwick. Not in Isle of Man and very local in Ireland, recorded from Cos Wexford, Dublin and Down only. North and central Palaearctic region to Japan; Iceland; North America.*

2 Size larger, 5.5-7.3 mm; elytra longer in proportion to width, about 1.6 x as long as broad; dorsum more thickly and uniformly clothed with linear, golden, recumbent setae or scale-like pubescence, forming obscure mottling at sides of elytra and without bare patches laterad, disc of first interstice not bare; sides of pronotum rounded but strongly convergent anteriad, distal width much narrower than basal (fig. 65)..*scirpi*

Fig. 65

Male antennae inserted within anterior third of rostrum. Female antennae inserted at one third from rostral apex, or a little further basad.

In marshes, fens and other wetlands, and at the sides of rivers, canals and other water courses. Biology little-known in Britain, but associated with species of Carex *and* Typha *here, and on the Continent also with* Scirpus *(in the broad sense; the genus has been split (Stace, 1991)). Larvae in rootstocks. Local and not generally common, though often in numbers when found. Widely distributed throughout England and Wales to Cumberland, but not recorded from Scotland or Isle of Man. Local, but widely distributed in Ireland. Throughout Europe and the Palaearctic region generally to Japan.*

- Size smaller, 3.4-5.1 mm; elytra shorter in proportion to width, about 1.4 x as long as broad; dorsum much less thickly and uniformly clothed with generally finer setae, elytra with bare, glabrous patches laterad, disc of first interstice devoid of setae, or with at most a few sparse, very small and fine setae; sides of pronotum more evenly rounded, distal width little narrower than basal (fig. 66). .... ***acridulus***

Fig. 66

Male rostrum a little more strongly punctured and duller, antennae inserted within the anterior third (fig. 67). Female rostrum a little less strongly punctured, more strongly shining and with antennae inserted at about one third from apex (fig. 68).

In fens, marshes and all kinds of damp and wet places. Also at the sides of rivers, streams and canals and water courses generally. Hosts and biology little-studied in Britain, but probably associated with semi-aquatic grasses, as in continental Europe where Glyceria maxima *is a common host. Larvae attack underground stolons. Common (though seldom taken in large numbers) throughout the British Isles from W Cornwall to E and W Sutherland, including all the Welsh vice-counties, and from S Co. Kerry to Co. Antrim. Recorded from Isle of Man and most of the Scottish island groups, including the Outer Hebrides, Orkney and Shetland. Throughout Europe, including Iceland, and the Palaearctic to the Pacific.*

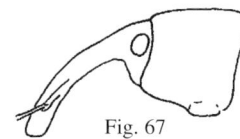

Fig. 67

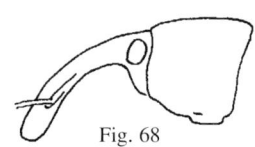

Fig. 68

Genus **Thryogenes** Bedel

Five species of this small genus have been described from the Palaearctic region, but *T. atrirostris* is regarded as conspecific with *T. fiorii*, following Booth (in press). The four species occurring in the British Isles include *T. fiorii*, which was found to be a longstanding resident in the British Isles (Booth, 1993).

Very little is known about the biology of species of *Thryogenes* in the British Isles. Continental work suggests that each species is associated with a different genus of wetland plants.

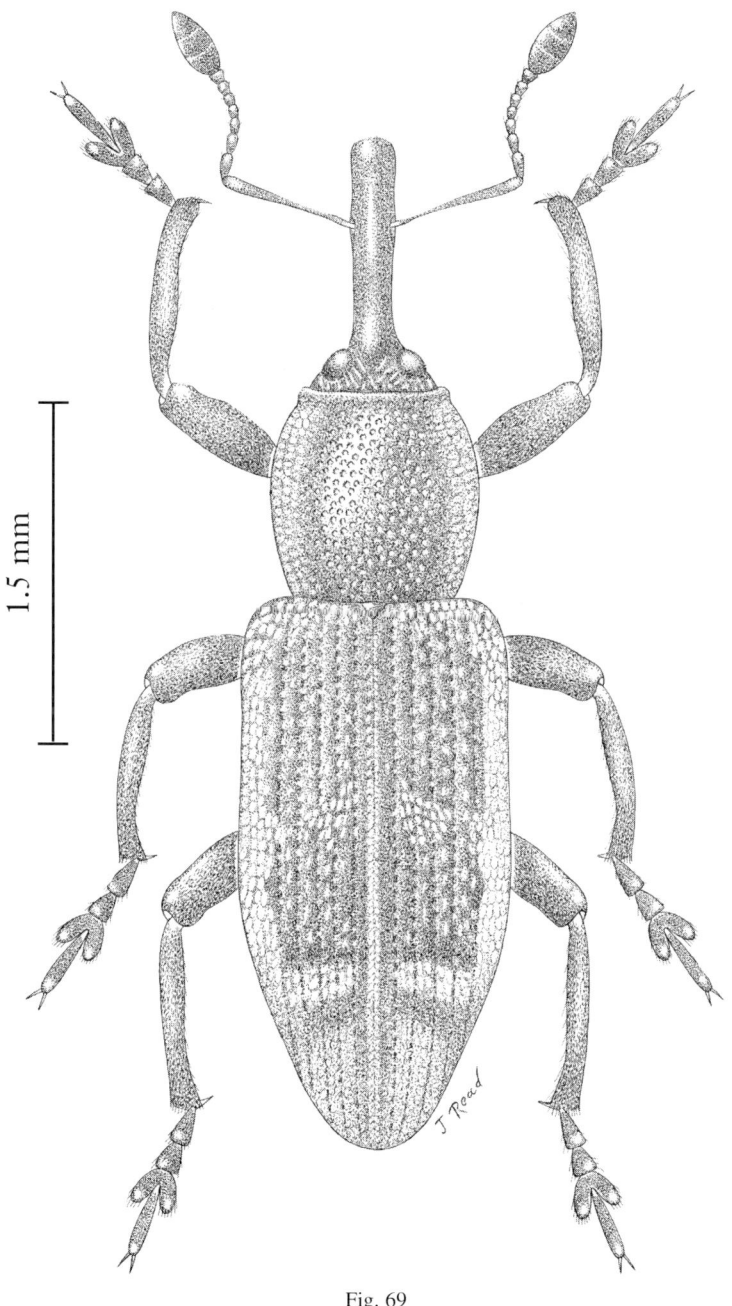

1.5 mm

Fig. 69
Thryogenes nereis

Key to species of *Thryogenes*

1 Narrower, elytra 1.85-2.00 x as long as broad, narrower in proportion to pronotum, 1.22-1.30 x its width; elytra straight to slightly convex at sides apicad (fig. 70). **2**

- Broader, elytra 1.70-1.80 x as long as broad, broader in proportion to pronotum, 1.36-1.46 x its width; elytra slightly concave at sides apicad (fig. 71)............................... **3**

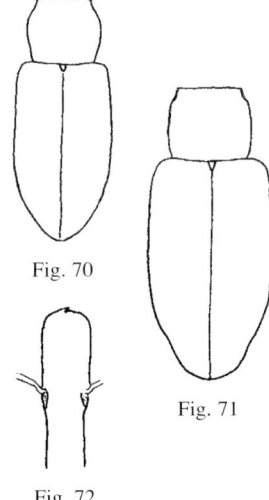

Fig. 70

Fig. 71

2 Rostrum red-brown, more strongly curved and shorter, less strongly expanded apicad (particularly in male, fig. 72); antennal scape shorter, clearly shorter than total length of fore-tarsus; eye in lateral view closer to top of head (vertex about level) (fig. 73) on average smaller, 2.9-4.2 mm] (habitus fig. 69).. ***nereis***

Fig. 72

Male rostrum more strongly curved, duller and shorter (not as long as width across elytra), more clearly dilated at apex; antennae inserted nearer rostral apex. Female rostrum less curved, more shining and longer (as long as elytral width), less strongly dilated at apex; antennae inserted further from apex of rostrum.

In fens, bogs, mires and at the sides of water-bodies. Foodplants in the British Isles uncertain, but almost certainly associated with Eleocharis palustris, *as in continental Europe. Also recorded from* Scirpus *spp.,* Juncus *spp. and other wetland plants, but evidence for these as hosts is lacking. Larvae in lower parts of the stem of* Eleocharis *(Dieckmann, 1986). Local, but not uncommon and sometimes abundant where it occurs. Throughout England and Wales as far north as Chester and NE York, but recorded only from Kirkcudbright in Scotland. Widespread and fairly common in Ireland; not recorded from Isle of Man. Throughout northern and central Europe.*

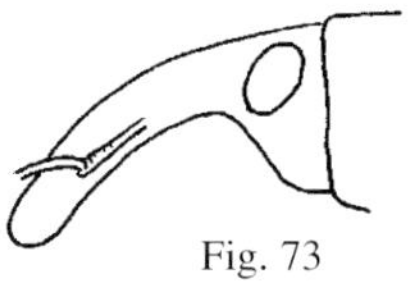

Fig. 73

- Rostrum dark brown to black, straighter and longer, more strongly expanded apicad (particularly in male, fig. 74); scape longer, equal in length to total length of fore-tarsus; eye in lateral view closer to middle of head (vertex sloping downwards) (fig. 75) [on average larger, 3.5-4.3 mm]. ... ***fiorii***

Fig. 74

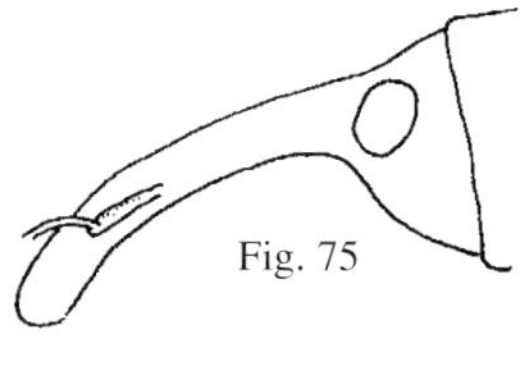

Male rostrum shorter, more strongly curved, more strongly punctured and duller; antennal insertion closer to rostral apex, scape considerably longer than apical part of rostrum (from antennal insertion). Female rostrum longer, less strongly curved, more finely punctured and more shining; antennal insertion further from rostral apex, scape very little longer than apical part of rostrum, subequal to it.

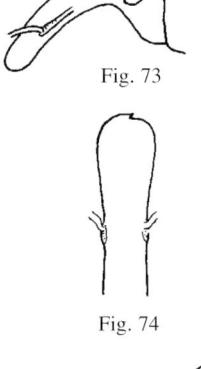

Fig. 75

In wet meadows and at the sides of water-bodies. An overlooked species in the British Isles, first recognised in the early 1990s (Booth, 1993), but known to have been collected in 1864 as well as since. Biology unknown in Britain although some specimens have been found consistently on Carex elata *(Morris, 2000). Associated with* Carex paniculata *in Germany (Dieckmann 1986); full range of hosts unknown. Larvae in the stems. Known from E Norfolk (several localities in the Broads), E Suffolk, Huntingdon (Woodwalton Fen) and W Sussex but likely to be more widespread. Southern and central Europe northwards to Denmark, but not known from Fennoscandia.*

3 Elytral vesture denser and more uniform, consisting of copious apressed isodiametric scales interspersed with fewer long-oval, golden scales and sparse elongate setae; elytra without a pattern; antennal funiculus longer, third joint about twice as long as broad, distinctly longer than fourth; last (seventh) joint quadrate to slightly elongate (fig. 76) [3.8-4.6 mm]... *festucae*

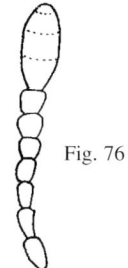

Fig. 76

Male rostrum shorter (about as long as head and thorax together), thicker, more strongly punctured and duller; antennae inserted at one-third from rostral apex. Female rostrum longer (longer than head and thorax together), thinner, less strongly punctured and more shining; antennae inserted a little further from rostral apex.

In wetlands, marshes and fens, and at the sides of rivers and other water-bodies. Hosts not well-known in the British Isles, but recorded in continental Europe in association with species of Scirpus *(s. l.; see under* Notaris scirpi *above). Larvae in the stems. Local and not common, but widely distributed in England and Wales from N Somerset eastwards to W Kent and northwards to Merioneth, Warwick and Leicester. Not recorded from south-west or northern England nor from Scotland or Isle of Man. Rare and very local in Ireland. Throughout central, southern and northern Europe; recorded from all the countries of Fennoscandia (which makes its apparent absence from northern Britain puzzling).*

-   Elytral vesture less dense and less uniform, consisting of fewer long-oval golden scales and copious long scale-like setae; isodiametric scales absent except along suture, so giving elytra a clearer pattern, accentuated by frequent patch of scales on each elytron behind middle; antennal funiculus shorter, third joint about 1.5 x as long as broad and not longer than fourth; last joint transverse (fig. 77) [3.30-4.35 mm]. .... *schirrosus*

Fig. 77

Male rostrum shorter, about as long as head and pronotum combined; antennae inserted closer to rostral apex, within apical third. Female rostrum longer, longer than head and pronotum combined; antennae inserted nearer rostral base, at a point one-third from apex.

In fens, marshes and wetlands generally. Hosts not well-known in Britain, but associated with species of Sparganium *both here (Joy, 1932) and in continental Europe; especially on* S. erectum. *Larvae almost certainly in stems. Rather rare, and with only sporadic records. In England from N Somerset eastwards to W Kent and northwards to Warwick and Salop with outlying records from Leicester and SE York. Not reported from Wales, Scotland or Isle of Man; Irish records are erroneous. Widely distributed in central and northern Europe.*

Genus *Grypus* Germar

Six species of this distinctive Holarctic genus are known, with two occurring in continental Europe, but only one has been recorded in the Britsh Isles.

-   One species; a dark weevil with a characteristic whitish elytral apex; less depressed than other erirhinines [4.1-6.7 mm].. *equiseti*

Sexes similar.

In fens, mires and wet places, including undercliffs with impeded drainage; also in wet and damp meadows and other grasslands. On species of Equisetum, *apparently only* E. arvense *and* E. palustre *(Cawthra, 1957b). Larvae in the stems; their morphology, and aspects of the species' biology, were described by Cawthra (1957a). Local, but not uncommon, and widely distributed throughout the British Isles as far north as Caithness, though with few records from northern Scotland and not reported from the Hebrides, Orkney or Shetland. Widely distributed in Ireland, but not recorded from Isle of Man. Widespread throughout the Palaearctic region; North America.*

Subfamily Lixinae

Although only a very few of the many continental European species of this extensive subfamily are found in the British Isles, we have representatives of the three tribes recognised as comprising the subfamily in western Europe. Several species have been included in the British list on very few, ancient records while the status of others appears to be, at best, dubious. On the other hand, one or two species have clearly declined spectacularly in historic time. The group includes some of our largest weevil species, though whether size itself has contributed to their vulnerability is debatable.

Secondary differences between the sexes are not marked in the subfamily, particularly in Cleonini and Rhinocyllini, though males often have a shallow depression on the ventral surface of the metathorax and abdomen. And, as in many weevils, the males are often slightly smaller and less robust, on average, than the females.

Key to tribes and genera

1 Rostrum terete, subcylindrical and smooth, not keeled at the sides (fig. 78), with, at most, a fine median dorsal keel; scrobe not continued beyond the antennal insertion anteriad (fig. 78).. **Lixini 3**

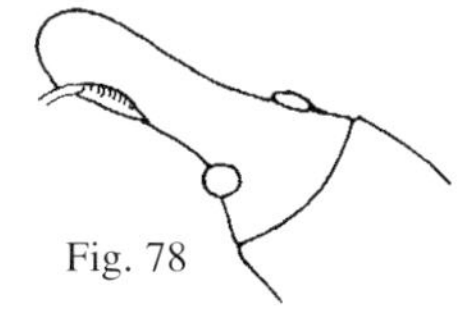

Fig. 78

- Rostrum strongly keeled, either at side or in the middle or both (fig. 79), not subcylindrical; scrobe continued anteriad beyond the antennal insertion (fig. 79). **2**

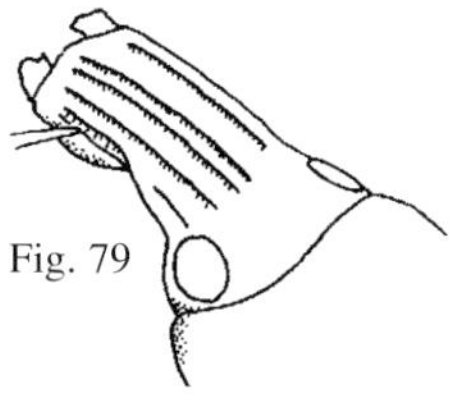

Fig. 79

2 Rostrum short, only just longer than broad, about as long as the head (fig. 103); second joint of hind tarsus shorter than third (fig. 80); [length not more than 7.0 mm]. .... **Rhinocyllini** (one genus, *Rhinocyllus*)

Fig. 80

- Rostrum long, at least twice as long as broad, much longer than head; second joint of hind tarsus longer than third (fig. 81); [length usually more than 7.0 mm]. .... **Cleonini 4**

Fig. 81

3 Elytra elongate, at least twice as long as broad; pronotum elongate.. *Lixus*

- Elytra broader, not more than 1.7 x as long as broad; pronotum strongly transverse (fig. 82)........................ *Larinus*

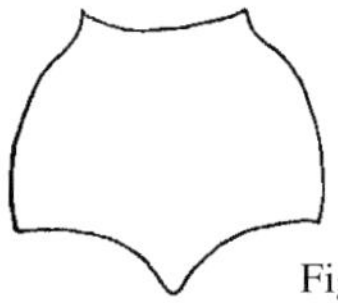

Fig. 82

4 First and second segments of hind tarsus more elongate, the second much longer (nearly x 2) than broad (figs 83, 84), the first with conspicuous, lateral, semi-recumbent setae; rostrum with a single ridge or keel which does not bifurcate anteriad. .... *Coniocleonus*

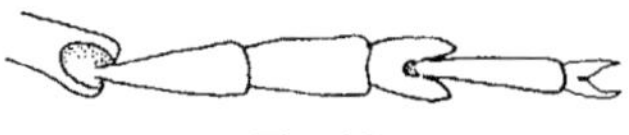

Fig. 83

- First and second segments of hind tarsus less elongate; the second not, or only slightly, longer than broad (figs 81, 88), the first without a trace of lateral setae; rostrum with either two median ridges or keels throughout (fig. 79) or a median keel which bifurcates just before the antennal insertion. **5**

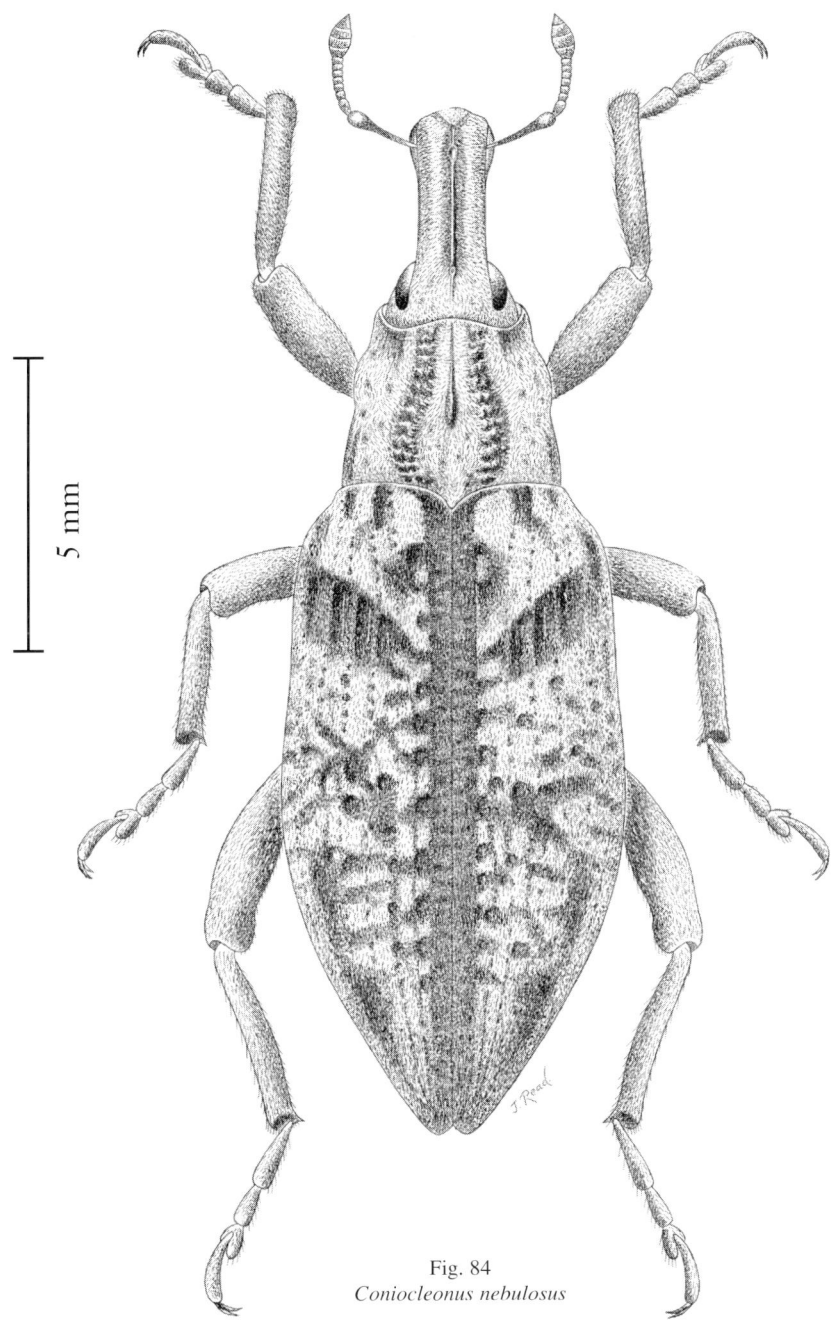

5 mm

Fig. 84
Coniocleonus nebulosus

43

5 Second segment of antennal funiculus clearly elongate (more than twice as long as broad), much longer, and clearly narrower, than first segment, club broader and more clearly demarcated from funiculus (fig. 85); size smaller, 6.0-11.0 mm; median keel of rostrum bifurcate anteriad. .... ***Bothynoderes***

\- Second segment of antennal funiculus at most only slightly longer than broad, shorter and only slightly narrower than first segment, club narrower, less clearly demarcated from funiculus (fig. 86); size larger, 10.0-16.0 mm; median keel of rostrum double throughout.. ***Cleonis***

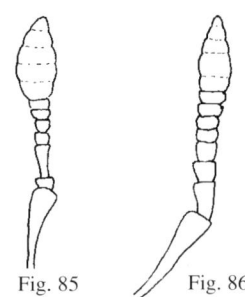

Fig. 85   Fig. 86

Genus *Coniocleonus* Motschulsky

About 20 species of this genus are known, most of them occurring in the Mediterranean region. Five species are quite widely distributed in central Europe but only two have been recorded from the British Isles, and one of these has long been extinct. However, as both species inhabit the same biotopes, and could be confused, it is desirable to include both here.

Key to species

Prothorax with a distinct tubercle or process, bluntly pointed, in front of each fore-coxa, readily seen in side or oblique view (fig. 87); elytra with two, not very distinct, bare, matt, diagonal markings, extending from the second interstice to about the sixth, the sides of these markings approximately straight; slightly more delicate and less robust species; length 11.5-14.5 mm (habitus fig. 84).
.... ***nebulosus***

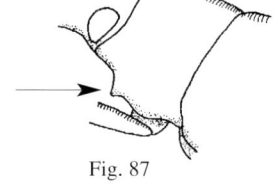

Fig. 87

Differences between sexes slight.
 On heathland and sandy places dominated by Ericaceae, often with areas of bare sand. In association with Calluna vulgaris, *the larvae in the stems and rootstocks. The larvae probably require old heather bushes in which to develop successfully, and the species possibly does not survive in heathland which is intensively managed. Very local, though widely distributed in southern England from W Cornwall (Hyman & Parsons, 1992) to Dorset, S Hants. and East Anglia northwards to Stafford, Derby and N Lincoln, but with few recent records. Not known from Wales, Scotland or Ireland. Throughout Europe, but generally rare.*

\- Prothorax with at most a weak, slightly raised area in front of each fore-coxa; elytra with, or without, two bare, matt markings which are curved and lie almost perpendicular to the suture at their inner ends; rather more robust, broader and less delicate species; length 9.0-13.5 mm.
.... ***hollbergi***

Differences between sexes slight.
 On heathland and, in continental Europe, open pine woodland. Foodplants uncertain; old records suggesting Pinus *(roots) were regarded as implausible by Dieckmann (1983), who suggested* Rumex acetosella; *though this is a small plant for so large a weevil; however, there is no strong evidence for* Calluna *(Hyman & Parsons, 1992). Extinct in Britain, and perhaps of uncertain provenance. Surrey (seven specimens), last dated record 1815 (Stephens, 1831). Europe to Siberia, widely distributed.*

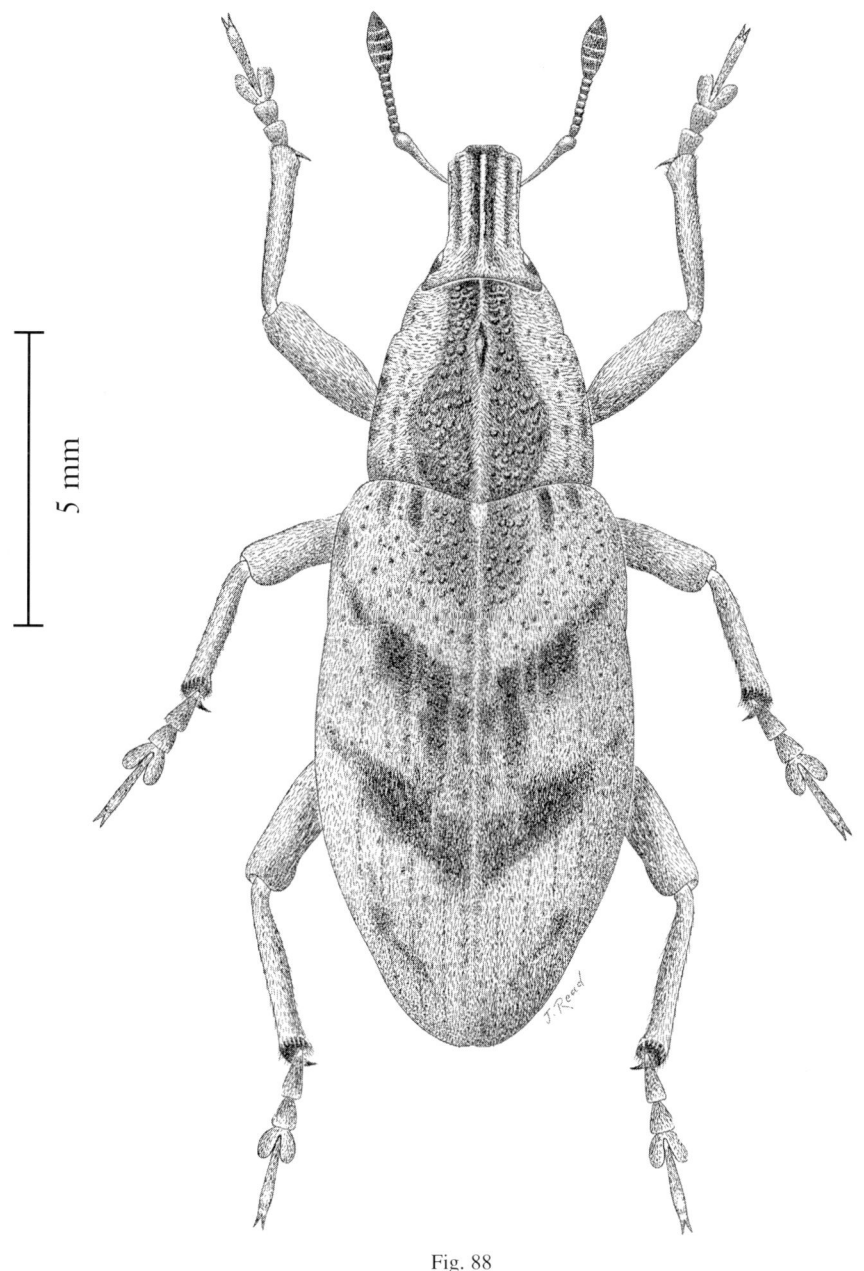

5 mm

Fig. 88
Cleonis pigra

Genus *Bothynoderes* Schoenherr

This genus contains only two species, one of which has been reported occasionally from Britain.

- One species: a relatively small cleonine (6.0-11.0 mm), with conspicuous pale, whitish pubescence interspersed with bare, black areas, often including a transverse band just behind the middle of the elytra.................................. ***affinis***

Sexual differences slight.

On sandy soils and in waste places (on the continent). Most British specimens have been taken near the coast. Associated with a wide range of Chenopodiaceae in continental Europe, including species of Chenopodium, Atriplex, Salsola *and* Beta, *the larvae in long-oval galls in the upper parts of the taproots; pupae are formed in the galls. Possibly an occasional immigrant to Britain (the adult weevils are fully winged); certainly no persistent colony is known. Recorded from Colchester (N Essex), Ipswich (E Suffolk), Thetford (W Norfolk), Cromer (E Norfolk) and near Edinburgh in the 19th century. Last record from the "outskirts of the New Forest" in 1936 (Dumper, 1937). No Welsh or Irish records. RDB3 (Hyman & Parsons, 1992, designate it as 'extinct' but have overlooked the most recent record). Europe, Asia Minor and central Asia to Siberia.*

Genus *Cleonis* Dejean

Only three species are currently referred to this genus, more familiar in older literature as *Cleonus* Schoenherr. This name is masculine, but *Cleonis* is feminine (Dejean, 1821), hence the change to the species-group name.

- One species; a large cleonine with very variable mottled markings, general facies grey to grey-brown. Length 9.0-14.8 mm (habitus fig. 88). ....***pigra***

Male abdomen impressed, apex subtruncate. Female abdomen simple, apex rounded.

*On sand dunes and other open, sandy places and so most usually on the coast, but recorded from a few inland localities, for example the Suffolk Breckland (Eversham & Telfer, 1995). On thistles (*Cirsium *and* Carduus *spp.) and other Asteraceae; recorded from a wide range of species in continental Europe, but most usually associated with* Cirsium arvense *in Britain. Larvae in the stems; aspects of the biology described by Cawthra (1958). Local, but sometimes abundant, recorded from most of the maritime vice-counties of England and from Glamorgan, Pembroke and Anglesey in Wales. On the south-east coast only in Scotland from Berwick to Forfar. Rare in Ireland (Co. Down only). Widely distributed throughout the Palaearctic region. Introduced into North America.*

Genus *Lixus* Fabricius

This is a very speciose genus of which a few of the many European species extend (or extended) as far as the British Isles. The elongate shape is related to the stem-boring habits of the larvae. Six species are currently included on the British list, but two are of ancient or dubious provenance and three are almost certainly or possibly extinct, leaving only one (a recent discovery) definitely extant. Species of *Lixus* (and *Larinus*) are notable for the pruina which covers fresh specimens in life. This is fugitive and seldom fully present in museum specimens.

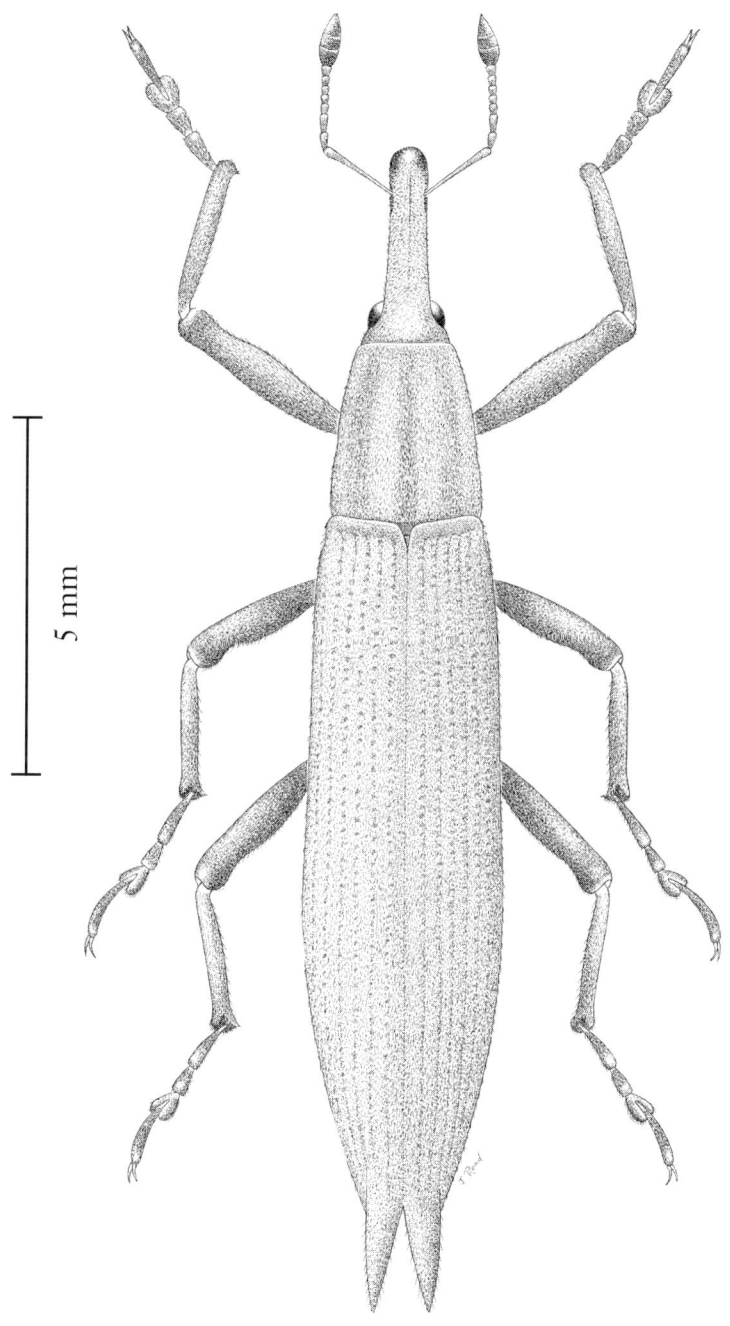

5 mm

Fig. 89
Lixus paraplecticus

Key to species

1 Scape longer than the width of rostrum at antennal insertion and longer than first four segments of funiculus taken together (fig. 90); larger species, length 8.0-17.5 mm.
.... **2**

- Scape shorter than width of rostrum at antennal insertion and not, or scarcely, longer than first three segments of funiculus taken together (fig. 91); smaller species, length 4.0-9.5 mm. **5**

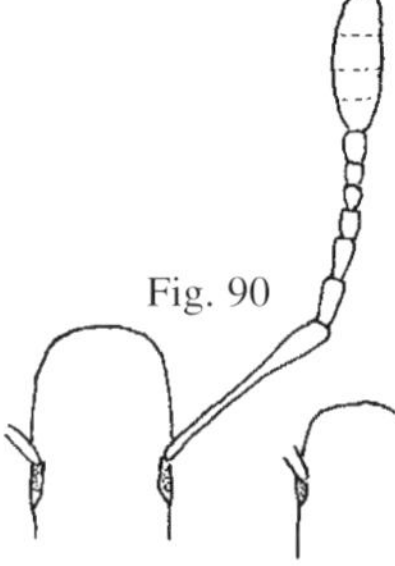

Fig. 90

Fig. 91

2 Each elytron sinuate at apex, prolonged into a tooth, or process (figs 92, 94-96), which is as long as, or longer than, first segment of hind-tarsus ; more elongate and delicate species [on Apiaceae, generally in damp situations or at watersides]. **3**

Fig. 92

- Each elytron rounded or subtruncate at apex (fig. 93), or with a short, blunt process, elytra not sinuate (figs 97, 101); more robust and less delicate species [not on Apiaceae, more clearly terrestrial]. **4**

Fig. 93

3 Elytral process longer and sharper (fig. 92), not upturned at apex (lateral view, fig. 94); eyes more rounded and prominent; legs longer and more delicate, fore-tibia as long as, or longer than, rostrum [length 10.0-17.0 mm] (habitus fig. 89). .... ***paraplecticus***

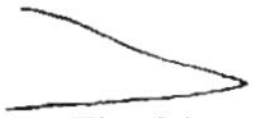

Fig. 94

Male rostrum slightly shorter, not as long as the pronotum, antennae inserted at about 1.5 x rostrum width from apex. Female rostrum slightly longer, as long, or nearly as long, as pronotum, antennae inserted at about 2 x rostrum width from apex.

In fens and wet places and on the banks of large rivers. On Sium latifolium, Oenanthe aquatica *and possibly other* Oenanthe *species in Britain, but recorded also from species of* Berula, Apium *and* Anthriscus *in central Europe. Larvae in the stems. Formerly widespread, though local, in England eastwards from N Somerset and the south coast northwards to Cumberland. No Welsh, Irish or Scottish records. Possibly extinct, last recorded from N Somerset in 1958 (Duff, 1993). It was common in the East Anglian fens during the early 19th century and on banks of the Thames at Fulham and Barnes (Stephens, 1831). It occurred plentifully along the banks of the Medway during the late 1940s and 1950s (Massee, 1940; 1963). It has disappeared from these sites because of land-use changes. RDB1. Widespread throughout Europe and also in the eastern Palaearctic.*

- Elytral process shorter and blunter (fig. 95), usually upturned at apex (lateral view; fig. 96); eyes almost flat; legs shorter and more robust, fore-tibia clearly shorter than rostrum [length 11.0-17.0 mm]. ....***iridis***

Fig. 95    Fig. 96

Male rostrum shorter and broader, shorter than head and pronotum and head together, apical part from insertion of antennae shorter than scape. Female rostrum longer and narrower, longer than head and pronotum together, apical part from insertion of antennae longer than scape.

Usually (in continental Europe) in marshes and other wet places. On a very wide range of Apiaceae (Dieckmann, 1983), the larvae in the stems. Doubtfully British and, if so, long extinct. A specimen in the Power collection was apparently taken in 1836 at either Hornsea Fen or Mildenhall (Fowler, 1891). In view of the rapid decline of other species of Lixus *in Britain, it is possible that it was once indigenous. Widely distributed in the Palaearctic.*

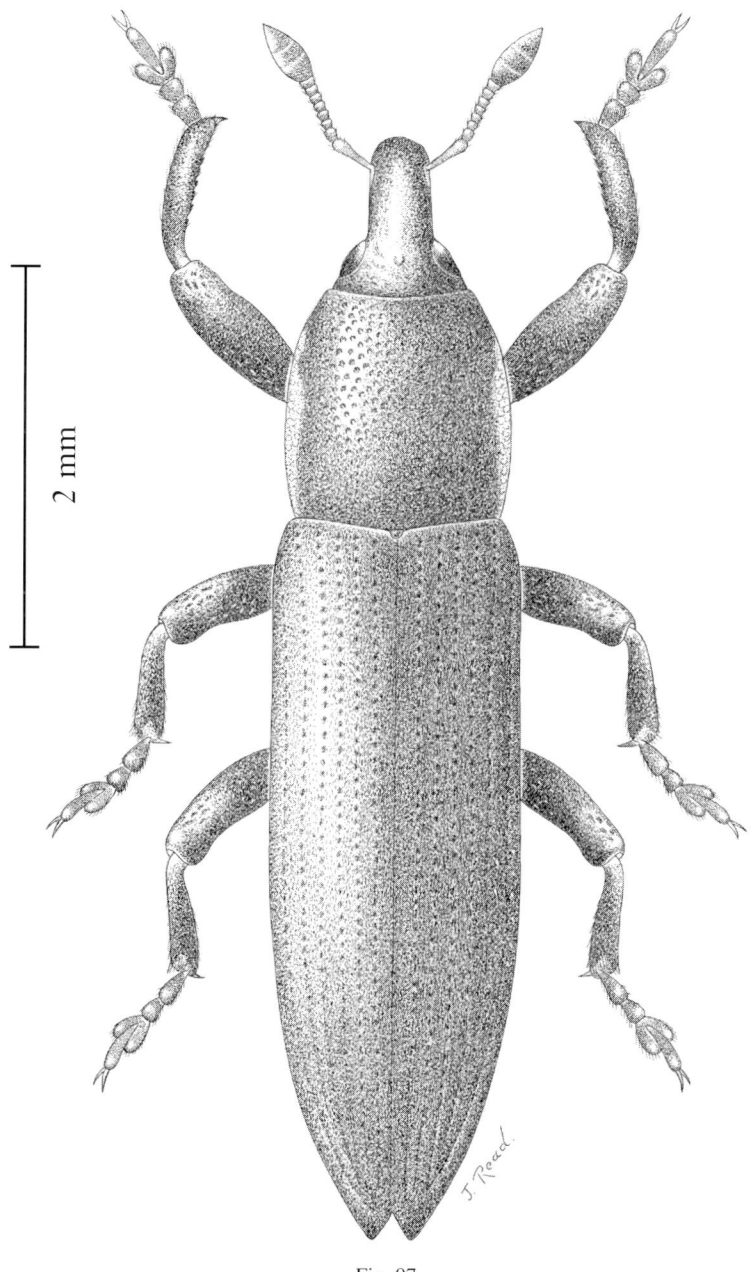

Fig. 97
Lixus scabricollis

4 Pronotum with a pale (whitish or yellowish) lateral band; second segment of funiculus much less than twice as long as third; pronotum more finely rugose; rostrum almost straight [elytra not produced at apex (fig. 98); length 5.5- 14.0 mm].. **vilis**

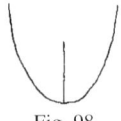

Fig. 98

Male antennal insertion at about one-third from apex. Female antennae inserted just in front of middle of rostrum.

On fixed, stable sand dunes and so usually on the coast. On Erodium cicutarium, *larvae in the stems and rootstocks and so requiring large, well-grown plants. British specimens appear to be mostly smaller than continental examples. Probably extinct in the British Isles. Formerly in S Somerset, S Hants. and, particularly, E Kent. Last record 1905. RDB1. Widely distributed in the Mediterranean region, central and Eastern Europe.*

- Pronotum without a pale, lateral band; second segment of funiculus twice (or more) as long as third; pronotum more coarsely rugose; rostrum distinctly curved [elytra not produced at apex (fig. 99); length 10.5-17.5 mm]. .... *angustatus*

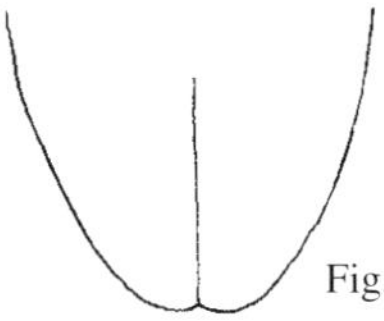

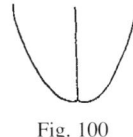

Fig. 99

Male rostrum shorter and more robust, antennae inserted about one-third from apex. Female rostrum longer and narrower, antennae inserted just in front of middle.

*In wet grasslands and other marshy places in the British Isles but not restricted to such places elsewhere. Predominantly on thistles (*Cirsium *and* Carduus *species) in Britain. On a wider range of foodplants abroad, including* Malva sylvestris *and* Vicia faba *(Dieckmann, 1983); a pest of beans in the Mediterranean region (e.g. Liotta, 1963). Probably extinct in the British Isles. Recorded from E and W Kent and W Sussex, but known mainly from the Fairlight and Hastings district of E Sussex, where J.A. Power took 150 specimens in a fortnight in 1867 (Fowler, 1891). Last record, in 1923, from Fairlight (Mitford, 1923). RDB1. Widely distributed throughout Europe and in Madeira and the Canaries, extending to central Asia.*

5 Rostrum longer, about as long as pronotum; head narrower, quadrate, sides convergent anteriad; eyes protuberant, remote from anterior margin of pronotum by more than width of eyes; pronotum transverse, sides strongly convergent anteriad; base of elytra strongly sinuate; apex of elytra very weakly, or not, produced (fig. 100) [length 4.0-9.5 mm, but very variable (in continental specimens)]. ....*elongatus*

Fig. 100

Male rostrum shorter and broader, duller, more strongly punctured. Female rostrum longer and more delicate, more clearly shining and less strongly punctured.

*Mainly in waste places and disturbed areas (in continental Europe). On a wide variety of thistles (*Carduus *and* Cirsium *species), larvae in the stems. Very doubtfully British. Recorded, with considerable circumstantial details, from near Devizes, N Wilts., in 1864 by Sidebotham (Rye, 1865; Fowler, 1891), but not found since. It is probable that some error was made, but the species is widely distributed in north-western Europe, though local (Dieckmann, 1983). Europe to central Asia, North Africa.*

- Rostrum short, clearly shorter than pronotum; head broader, transverse; eyes flat, close to anterior margin of pronotum, not separated from it by more than width of eye; pronotum quadrate, its sides subparallel; base of elytra straight, elytral apices clearly produced (fig. 101). Length 4.5-6.0 mm (habitus fig. 97). ....*scabricollis*

Fig. 101

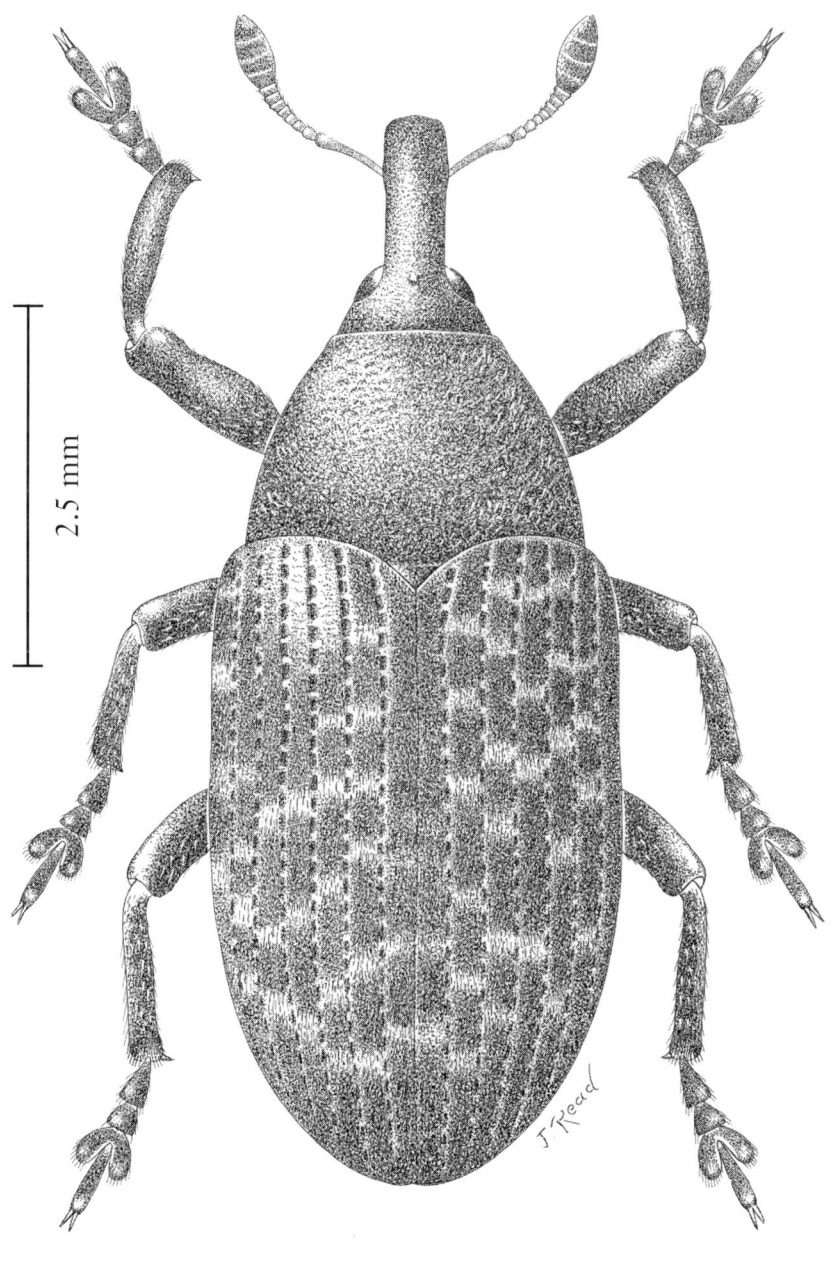

2.5 mm

Fig. 102
Larinus planus

Male rostrum slightly shorter, more coarsely punctured and less clearly shining; antennae inserted well in front of middle. Female rostrum longer, more finely punctured and more strongly shining; antennae inserted only just in front of middle.

Mainly on shingle beaches in Britain, though in a much broader range of biotopes in continental Europe. On species of Chenopodiaceae, especially Beta *and* Atriplex, *the larvae in the stems. A recent discovery in Britain (Heal, 1992), possibly only recently introduced, but now well-established in several areas of southern England (W Kent, Sussex, Dorset) and South Wales. Widely distributed in southern Europe and N. Africa, but in central Europe recorded only from western Germany (as an introduction).*

Genus *Larinus* Germar

Like *Lixus*, this is a species-rich genus with many representatives in southern Europe and the Mediterranean region. Several species have been screened as possible biological control agents of thistles and other weed species of Asteraceae-Cynareae (Zwölfer, 1965). However, only a single species inhabits the British Isles.

- One species; entirely black, with sparse pubescence forming a tessellated pattern on the elytra, and with bright mustard-yellow pruina in life; length 4.8-9.5 mm (habitus fig. 102). .... *planus*

Male rostrum slightly broader, more robust, duller and more closely and rugosely sculptured, antennae inserted a little closer to apex, at less than the length of the antennal scape from its tip. Female rostrum slightly narrower, less robust, more shining and less rugosely sculptured, antennae inserted less close to apex, at a distance equivalent to about the length of the antennal scape.

*In grasslands and open places, often in disturbed areas and on the coast. On various species of thistles (*Cirsium *and* Carduus*) spp., possibly also on* Carlina vulgaris *and* Centaurea *spp. Larvae in the flower-heads. Widespread in southern English coastal counties from E Cornwall to E Kent, northwards to E Gloucester; also in Wales (Brecon,, Radnor, Cardigan, Montgomery and Merioneth). No records from northern England, Scotland or Ireland. Generally scarce, but sometimes locally abundant. Widespread in Europe, Asia Minor and North Africa.*

Genus *Rhinocyllus* Germar

The tribe Rhinocyllini is a small one, with only two genera, restricted to the Palaearctic. *Rhinocyllus* contains only six species, with only one widespread in Europe, including the British Isles.

- One species; entirely black, with pale pubescence forming a tessellated pattern on the elytra; rostrum very short, but otherwise superficially very similar to *Larinus planus*; length 4.2-6.7 mm (habitus fig. 103).............................. *conicus*

Male ventrally with a shallow, longitudinal median depression, extending from the metathorax to the second (visible) ventrite. Female without such a depression, ventral surface weakly convex.

*In open and disturbed places, usually on the coast. On thistles (*Cirsium *and* Carduus *spp.), the larvae in the flower-heads. Widespread, but local, in the southern coastal counties of England from W Cornwall to E Kent; other records, from inland counties, require confirmation. No Welsh, Scottish or Irish records. Widespread in southern and central Europe to North Africa and central Asia. Introduced into North America.*

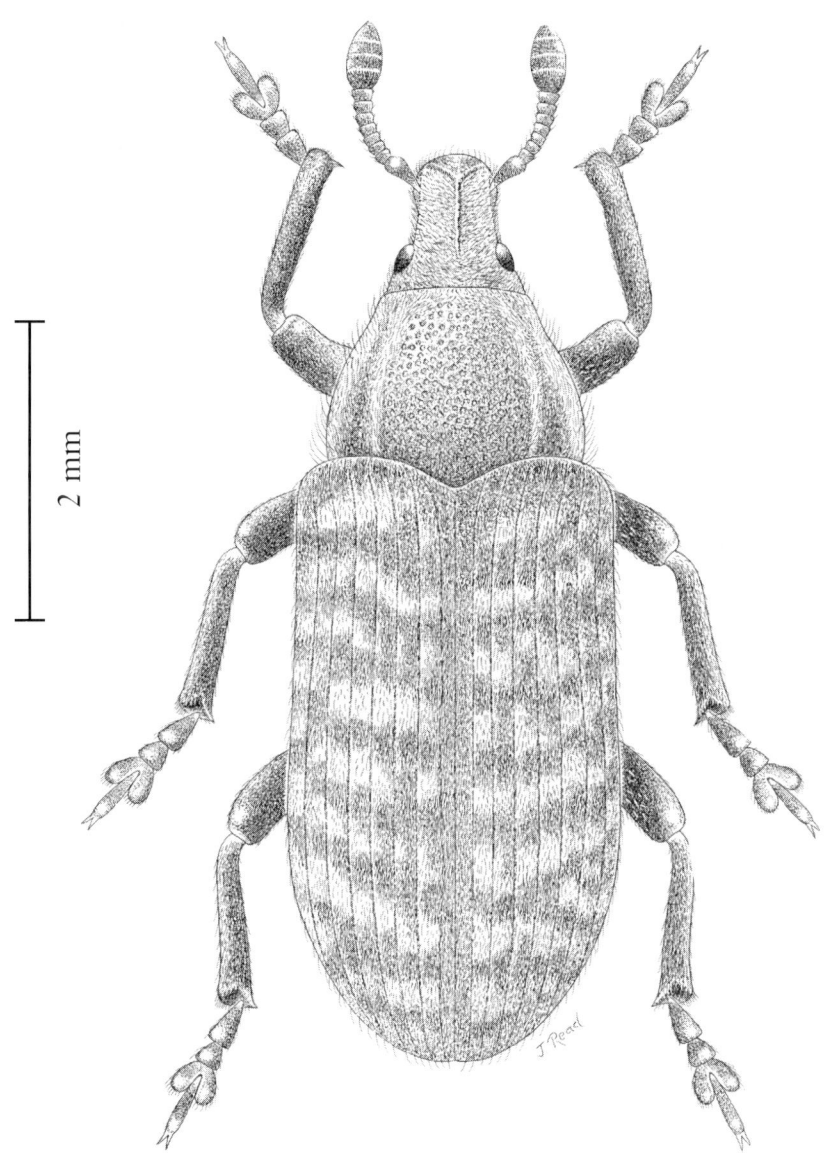

Fig. 103
Rhinocyllus conicus

Subfamily Hyperinae

About 30 genera and 400 species are known in this subfamily, which is cosmopolitan, though with most species inhabiting the Palaearctic region. Of the five genera recorded from western Europe, we have representatives of only two; no species of *Donus* (a genus of flightless species), *Pachypera* nor *Coniatus* (associated with Tamaricaceae) is known from the British Isles.

Biologically the subfamily is notable for having ectophytic, pigmented, actively mobile larvae which resemble small caterpillars. They pupate in cocoons which are of two types according to the species; 'latticed' cocoons are formed of a loose mesh of threads and are open to the air, whereas 'closed' cocoons are similar but lined with a complete film of the cocoon material and thus are not exposed to the air (Tempère, 1972). In most species, perhaps all, the pupal stage is short, lasting only about 7-10 days.

Key to genera

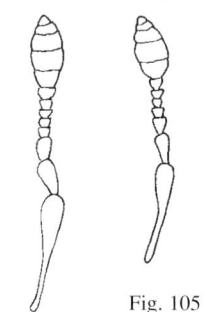

1 Antennal funiculus with seven segments (fig. 104); size generally larger, 3.5-10.0 mm [including common and widely distributed species; on a wide range of plant hosts, but including Geraniaceae].. *Hypera*

- Antennal funiculus with six segments (fig. 105); size small, 2.5-4.0 mm [local and rare species, on Geraniaceae only].
.... *Limobius*

Fig. 105

Fig. 104

Genus *Hypera* Germar

Sixteen species of this genus have been recorded from the British Isles, about twice this number from central Europe and over 120 from the Palaearctic region. The species were revised by Petri (1901). They include some of our commonest and most familiar weevils. The range of foodplants is wide, with species of Papilionaceae, Caryophyllaceae, Apiaceae, Polygonaceae and Geraniaceae being hosts to various species (Table 2). The larvae are easily reared (Dieckmann, 1989) and those of some species have been described and keyed (Anderson, 1948; Scherf, 1964; Dieckmann, 1989). Several *Hypera* are minor pests, particularly of legumes, in this country (Gratwick, 1992), but the importance of the *brunnipennis-postica* complex as pests of sweetclover (alfalfa), particularly in USA, has led to much work on its physiology, ecology and control. For another recent key to the species see Fowles (1994).

Key to species

Fig. 106 Fig. 107

1 Recumbent scales of elytra convex, truncate or at most weakly concave at apex (figs 106, 107). 2

- Recumbent scales of elytra bifid, deeply divided at apex, depth of the cleft at least as great as the distance between the two arms at their apices (fig. 108). 6

Fig. 108

2 Rostrum short, about 1.6-1.8 x as long as broad; pronotum broadest well in front of middle, at about 2/3 from base (fig. 109); larger and more robust species [on *Trifolium* spp.; predominantly ground-living], 6.8-8.5 mm........ *punctata*

Fig. 109

54

Male fore-tibia less robust, strongly curved inwards at apex, outer margin strongly curved, inner margin sinuate (fig. 110); elytra narrower, more nearly parallel-sided, broadest at about middle; all tibiae toothed at apex, mid-tibial tooth strong. Female fore-tibia more nearly straight, very weakly convex on outer margin, inner margin more weakly sinuate (fig. 111); only fore- and mid-tibiae toothed at apex, mid-tibial tooth weak; elytra broader, sides divergent so often broadest behind middle.

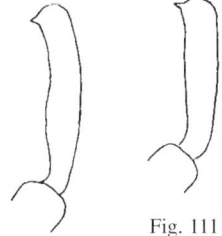

Fig. 111

Fig. 110

In grassland, croplands and on the coast, often in somewhat open situations and usually on the ground, seldom at any height on vegetation. On Trifolium *spp., especially* T. repens *and* T. pratense, *and possibly also on* Medicago *spp. (a minor crop pest in France). Larva green, with narrow, white, purple-edged midline; cocoon 'latticed', yellowish. There appear to be two independent generations, one with overwintering larvae the other with larvae in April-June, as in continental Europe (Scherf, 1964). Very widely distributed (though seldom abundant) throughout the British Isles, including the Outer Hebrides and Orkney, but not recorded from Shetland. Whole Palaearctic region. Introduced into North America.*

- Rostrum longer, at least twice as long as broad; pronotum broadest at or behind middle; smaller species [on Geraniaceae, Apiaceae or Polygonaceae, often on their aerial parts], 4.0-7.5 mm. **3**

3 Alternate elytral interstices strongly raised (all interstices convex), interstices with long, upstanding, conspicuous semi-erect, pointed setae arranged in a distinct longitudinal row on each; derm of tibiae reddish; pronotum strongly transverse (1.25-1.35 x as broad as long); rostrum shorter, about twice as long as broad [on *Erodium*], 4.9 - 6.5 mm. ....***dauci***

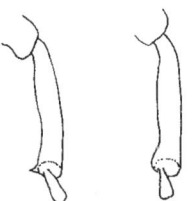

Fig. 112   Fig. 113

Male fore-tibia bent inwards at apex; mid-tibia with a strong apical tooth (fig. 112). Female fore-tibia almost straight at apex; mid-tibia untoothed (fig. 113).

In open, sandy places, usually by the sea but also inland, for example in the East Anglian Breckland. On Erodium cicutarium, *probably exclusively so in the British Isles and perhaps also in central Europe (Dieckmann, 1981, 1989), though reported in the older literature from other* Erodium *and* Geranium *spp. (Scherf, 1964). Larva green with a broad, longitudinal median white band, and similar to the caterpillar of the butterfly* Aricia agestis *(Newman, 1858); cocoon 'closed'. Local, and generally scarce, but widely distributed on the coasts of England and Wales; in Scotland as far north as Edinburgh and with an isolated inland record from W Perth. Recorded in Ireland only from W Co. Donegal. Widely distributed in central and southern Europe to the Caucasus and in North Africa.*

- All elytral interstices flat to weakly convex, alternate interstices not raised, each with at most short, more nearly recumbent setae; derm of tibiae black; pronotum less strongly transverse (about 1.10-1.20 x as long as broad); rostrum longer, 2.5-3.5 x as long as broad [on Apiaceae or Polygonaceae], 4.0- 7.5 mm. **4**

4 Shoulders inconspicuous, elytra scarcely broader than pronotum at base, its sides divergent to apex; elytra distinctively yellow-brown, with sparse, small, isodiametric dark patches on disc and a larger sutural macula at apex; each elytral interstice with many small, golden, semi-erect setae, disposed evenly but not forming distinct longitudinal rows [species probably extinct in the British Isles], 6.5-7.5 mm. .... ***arundinis***

55

Male fore-tibia bent inwards and downwards at apex; mid-tibia with a strong apical tooth. Female fore-tibia almost straight at apex; apical tooth of mid-tibia smaller.

In marshes and fens and on the banks of rivers. Larvae monophagous on Sium latifolium, *though the adults also feed on other Apiaceae. Very rare in England and almost certainly extinct. The records, dating from the early and mid 19th century, were summarised by Fowler (1891). Sium latifolium has not been recorded from some of the putative localities and has become extinct at others. Widely distributed in northern and central Europe.*

- Shoulders more marked, elytra clearly broader than pronotum at base; sides of elytra subparallel; elytra variegated in shades of brown to black, never unicolorous yellowish with small maculae; black marks, if present, more numerous and extensive; each interstice with at most a few semi-erect setae [common and widely distributed species], 4.0-7.0 mm. **5**

5 Apex of rostrum glabrous and shining; vertex with setae or narrow, acuminate scales; underside of head with setae or narrow scales; sutural striae strongly convergent at elytral apex, not united with tenth striae (fig. 114); elytral pattern very variable; often with alternate light and dark interstices (f. *alternans* Stephens), or interstices with dark, isodiametric spots, but third interstice almost always paler than second or fourth [on Apiaceae, usually in damp places], 4.9-5.8 mm... ***pollux***

Fig. 114

Male fore-tibia conspicuously and rather abruptly bent inwards at about 1/3 from apex; all tibiae sharply toothed at apex; rostrum shorter, about as long as the pronotum, antennae inserted nearer its apex, at about 1.0 x rostral width. Female fore-tibia weakly convex on outer margin, not conspicuously bent inwards at apex; tibiae more weakly toothed; rostrum longer, longer than pronotum, antennae inserted further from its apex, at about 1.5 x rostral width.

Usually in wet and marshy places and on the banks of ponds and watercourses. On various Apiaceae, including species of Apium, Peucedenum, Oenanthe *(especially* O. crocata*), and reputedly also* Daucus. Crithmum maritimum *has recently been confirmed as a larval host (Fowles & Hammett, 2001). Larva dirty yellowish-white with greenish, lateral, longitudinal bands; cocoon 'latticed'. Rather local, but widely distributed throughout England and Wales and extending to the Scottish borders (Dumfries, Kirkcudbright and Ayr) and with an outlying record from Westerness. Rare in Ireland. Throughout Europe to Siberia and Japan.*

- Apex of rostrum sparsely pubescent and dull; vertex with truncate scales; underside of head with broad scales; sutural striae divergent at elytral apex, united with tenth striae (fig. 115); pattern of elytra variable, often with an obscure, broad, pale mark on each elytron behind middle (best seen with the naked eye) [on Polygonaceae, often in drier areas, waste places, etc.], 4.0- 5.5 mm. .... ***rumicis***

Fig. 115

Male fore-tibia bent inwards at apex, external margin curved; apical tooth of all tibiae stronger; distance from antennal insertion to apex of rostrum less than its width. Female fore-tibia not bent inwards at apex, external margin almost straight; all tibiae less strongly toothed at apex; distance from antennal insertion to rostral apex slightly greater than its width.

*In open, often waste, places; on roadsides, field margins and on the coast. On various 'docks' (*Rumex *(*Rumex*) spp.), including* R. crispus, R. hydrolapathum, R. obtusifolius, *etc. In continental Europe also on species of* Oxyria, Polygonum *and* Rheum. *Larva dark green with paler longitudinal stripes in midline and at sides; cocoon 'latticed'. Generally common and widely distributed throughout England, Wales and Ireland, extending as far north as Forfar and N Ebudes in Scotland. All Europe to the Caucasus, North Africa. Introduced into N. America.*

6 Fore-tibia with a distinct tooth or pointed process on the inner side at about its mid-point (figs 116, 117); elytral interstices raised, more strongly in some places than others; third interstice conspicuously raised anteriad; elytra with short pale and dark longitudinal stripes, 4.9-6.1 mm. .... *arator*

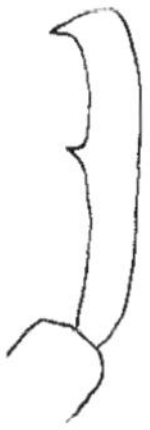

Fig. 116    Fig. 117

Male fore-tibia with a stronger and sharper median tooth (fig. 116); mid-tibia strongly toothed at apex; pronotum and elytra narrower. Female fore-tibia with a weaker and blunter median tooth (fig. 117); mid-tibia untoothed at apex; pronotum and elytra broader.

In dry, open places, often on light soils. Frequently near the coast on cliffs, shingle and the dryer parts of salt marshes, but not exclusively maritime. On a wide variety of Caryophyllaceae, including Stellaria, Cerastium, Silene, Spergula, Spergularia, *and* Lychnis *spp. (probably on most of the genera in the family with large-growing species). Larvae appear to be variable in colour, perhaps matching parts of the plant hosts other than the leaves, particularly the flowers. Cocoon 'closed'. Generally common, though only occasionally ocurring in numbers. Widely distributed throughout England, Wales, Ireland and the Isle of Man, and in Scotland as far north as Elgin and N Ebudes (not recorded from the northern mainland, Outer Hebrides, Orkney or Shetland). Throughout Europe to Siberia, North Africa.*

- Fore-tibia without a median tooth on the inner side; interstices flat to slightly convex, but not raised in some places; elytra generally without distinct longitudinal stripes, though if so also with dark isodiametric patches or other markings. **7**

7 Each elytral interstice with a row of long, conspicuous, erect setae, about as long as tibiae are wide; tibia with long, semi-erect setae; pronotum strongly transverse (about 1.5 x as broad as long), strongly contracted at base, sides slightly sinuate (fig. 118); elytra with alternate interstices silvery-white and grey, giving a distinctive striped appearance, though interstices also with black, or dark, isodiametric spots [rare and very local species, on *Daucus*], 4.4 - 5.2 mm. .... ***pastinacae***

Fig. 118

Male fore-tibia somewhat abruptly bent inwards about 2/5 from apex. Female fore-tibia evenly curved on the outer side.

On maritime cliffs in Britain, though not restricted to them in continental Europe. On Daucus carota *(a pest of cultivated carrot in France) and recorded from* Pastinaca sativa *elsewhere in Europe. Larva bright green with five whitish longitudinal stripes; it feeds mainly on the umbels of the host. Cocoon type not known. Very local and not common. Reliably recorded only from the Folkestone area of E Kent. RDB 1. Southern Europe and the Mediterranean region.*

- Elytra without very long, erect setae; tibiae without long, semi-erect setae; pronotum less strongly transverse (about 1.35 x as broad as long, or less); elytra without silvery-white and grey stripes [including common and widely distributed species, on Papilionaceae and Caryophyllaceae]. **8**

8 Eyes narrow-oval, 1.6-2.2 x as long as wide, seldom less than 1.8 x, flat (fig. 119); distance between eyes narrow, less than the width of an eye measured perpendicularly to the rostral mid-line [including very common species, on Papilionaceae].. **9**

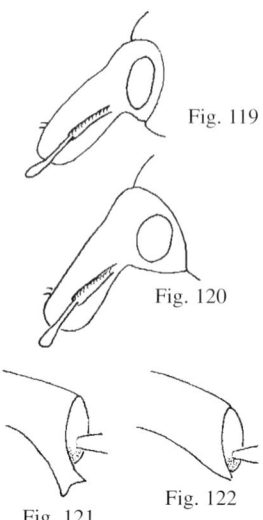

Fig. 119

- Eyes broad-oval, 1.3-1.4 x as long as wide, rounded (fig. 120); distance between eyes greater, about equal to the width of an eye measured perpendicularly to the rostral mid-line [pronotum without bifurcate scales on disc; rare species, on Caryophyllaceae], 5.2-5.6 mm. ..... *diversipunctata*

Fig. 120

Male fore-tibia less robust, strongly bent inwards at about 1/4 from apex, inner margin more strongly sinuate; all tibiae strongly toothed at apex, hind tibial tooth bicuspate (fig. 121). Female fore-tibia more robust, outer margin more regularly and gently curved, inner margin less strongly sinuate; all tibiae less strongly toothed at apex, hind tibial tooth simple (fig. 122).

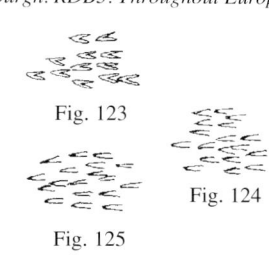

Fig. 122

Fig. 121

In a variety of open situations, including both dry and marshy areas; without a clear association with particular biotopes. On various Caryophyllaceae, including Cerastium arvense, Stellaria media, S. alsine *and* Myosoton aquaticum, *but with few foodplant records in Britain, so possibly also on other species. Larva pale green with a whitish median longitudinal line; cocoon 'closed'. Very scarce. A flightless species, widely, but discontinuously, distributed in central and northern England, Wales and the Scottish borders from Radnor, E and W Suffolk northwards to Dumfries, Selkirk and Edinburgh. RDB3. Throughout Europe, but generally rare.*

9 Elytral interstices with a longitudinal row of conspicuous, upstanding, semi-erect setae, particularly evident at declivity; cleft of elytral scales deep, apices of their arms long and tapering and often divergent (figs 123-125); disc of pronotum usually with some bifurcate scales [mostly small species, 3.2-5.3 mm (but *fuscocinerea* 5.8-6.7 mm)]. **10**

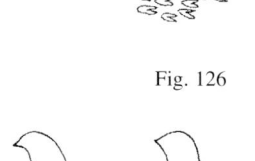

Fig. 123

Fig. 124

Fig. 125

- Elytral interstices with a longitudinal row of inconspicuous, short, almost recumbent setae; cleft of elytral scales shallow, less than half the scale-length, apices of their arms shorter, more nearly parallel-sided (fig. 126); disc of pronotum seldom with bifurcate scales [large species, 5.3 - 5.8 mm]. .... *suspiciosa*

Fig. 126

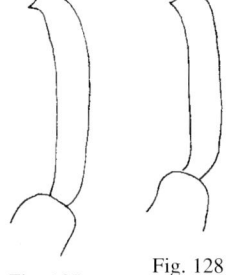

Male fore-tibia abruptly bent inwards at about 1/4 from apex (fig. 127); elytra conspicuously narrower and less rounded at sides, about 1.50-1.56 x as long as broad. Female fore-tibia bent inwards at apex, but outer margin forming a regular curve (fig. 128); elytra clearly broader and more rounded at sides, about 1.35-1.40 x as long as broad.

In waste places, on roadsides etc., in tall herb communities and in open situations with luxuriant herbaceous vegetation. On various Papilionaceae, including species of Lathyrus, Melilotus, Vicia, *especially* V. cracca, *and possibly* Trifolium. *Larva bright green with paler median longitudinal stripe; cocoon 'closed'. Local and infrequently found, but widely distributed throughout England and Wales and extending northwards to N Aberdeen, though with few Scottish records. Recorded in Ireland only from Co. Clare. Throughout Europe.*

Fig. 127

Fig. 128

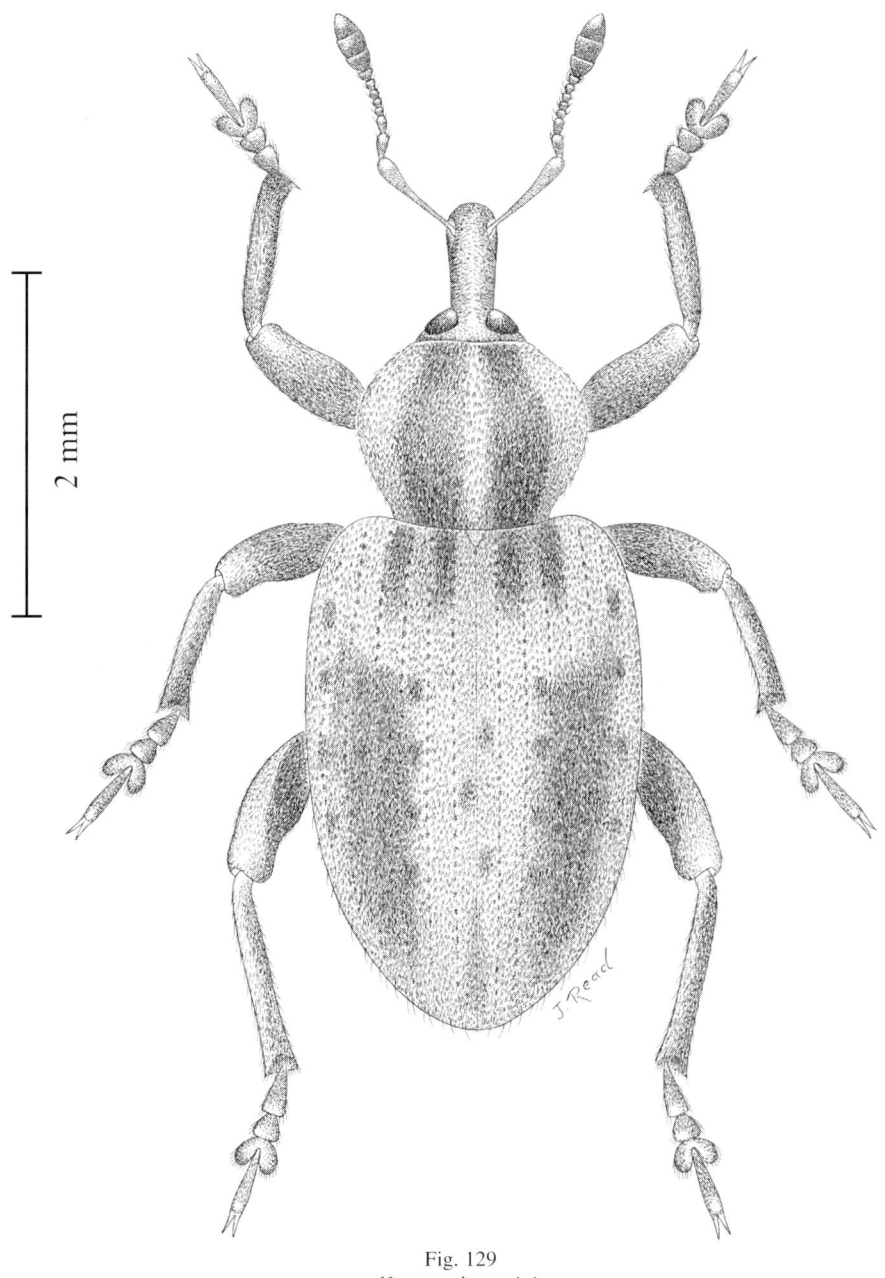

2 mm

Fig. 129
Hypera plantaginis

10 Elytral scales divided almost to base, sides almost straight throughout their length, and with no broad basal part to the scale (figs 130, 131); anterior surface of femora with setae or narrow, acuminate scales only; ventral surface of abdomen, from mid-coxae to apex, with setae or narrow acuminate scales only.. **13**

Fig. 130    Fig. 131

- Elytral scales not divided almost to base, cleft to middle or a little beyond it, sides more strongly curved, especially towards scale base, with a broad basal part to the scale (fig. 132); anterior surface of femora with mixed setae and bifurcate scales; ventral surface of abdomen, from mid-coxae to apex, with bifurcate scales, with or without setae... **11**

Fig. 132

11 Elytra with disc usually dark, the dark mark covering interstices 2 and 3 at base and extending further posteriad, at least to middle, without a conspicuous dark mark at sides covering interstices 4 to 6; pronotum less strongly rounded at sides and less strongly transverse, about 1.15-1.30 x as broad as long, broadest at or behind middle (fig. 133); median longitudinal line of pronotal scales less clear and narrower [male median lobe long, apical portion (apex to base of apodemes) 3.0 - 3.5 x as long as broad (fig. 138); on *Medicago*]..**12**

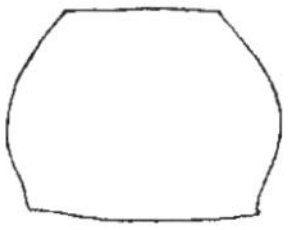

Fig. 133

- Elytra with disc pale, with a short dark mark covering interstices 2 and 3 at base but not extending to middle, but with a long, conspicuous dark mark at sides covering at least interstices 4 to 6 and at least as long as half the length of the elytra; pronotum strongly and evenly rounded at sides, more strongly transverse, about 1.27-1.36 x as broad as long, broadest slightly in front of middle (fig. 134); median longitudinal line of pronotal scales broad and distinct, nearly as wide as distance between eyes [male median lobe very short, apical portion (apex to base of apodemes) about 1.6 x as long as broad (fig. 135); on *Lotus*]; 4.0-4.8 mm (habitus fig. 129). .... ***plantaginis***

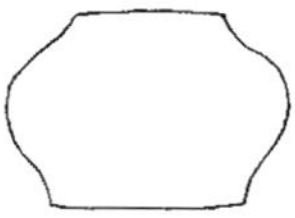

Fig. 134

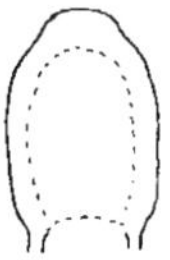

Fig. 135

Male fore-tibia narrower and more clearly bent inwards at apex; mid-tibia more strongly toothed at apex. Female fore-tibia broader and more evenly curved inwards at apex; mid-tibia less strongly toothed.

In grasslands, at roadsides and by the sea on cliffs and in open places. On species of Lotus, *primarily* L. corniculatus. *Larva green with a whitish, longitudinal median stripe, feeding on flowers as well as foliage; cocoon 'closed'. Common and very widely distributed throughout the British Isles. Recorded from most English and Welsh vice-counties; common, though less well- recorded in Ireland. Distributed throughout Scotland to the Outer Hebrides and Shetland, though not recorded from Orkney. Isle of Man. All Europe and the Mediterranean region.*

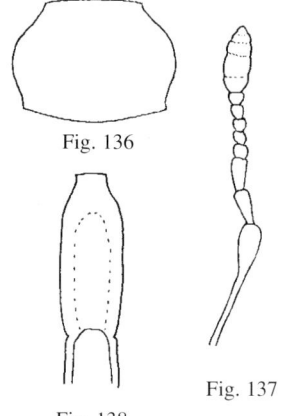

Fig. 136

Fig. 137

12 Larger, 5.8-6.7 mm; pronotum more strongly transverse, about 1.20-1.30 x as broad as long, more rounded at sides and clearly contracted to base (fig. 136); segments 5 and 6 of antennal funiculus quadrate, or nearly so (fig. 137) [uncommon and local; median lobe with a small subapical constriction (fig. 138)].. *fuscocinerea*

Fig. 138

Male fore-tibia more strongly bent inwards at apex, outer margin strongly curved (fig. 139). Female fore-tibia less clearly bent inwards at apex, outer margin less strongly curved (fig. 140).

In dry, open sandy and chalky situations and in grasslands. On species of Medicago, *including* M. sativa, M. falcata *and, probably,* M. lupulina. *In continental Europe also recorded from other species of Papilionaceae, mainly* Melilotus, Trifolium *and* Vicia *spp. Larva not described; cocoon 'latticed'. Scarce and usually rare, but widely distributed. In southern England northwards to NE York and in Scotland from the borders to W Perth, but not recorded from the northern mainland nor any of the Islands. Isle of Man. Very rare in Ireland. Europe and the Mediterranean region to Asia Minor.*

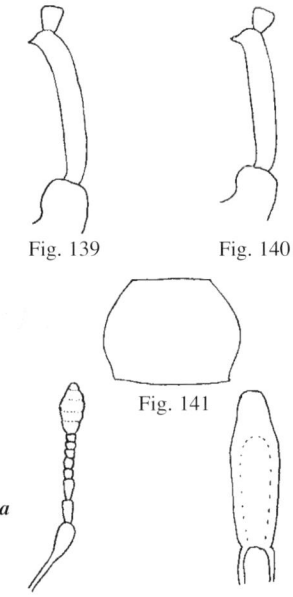

Fig. 139   Fig. 140

Fig. 141

- Smaller, 3.9-5.3 mm; pronotum less strongly transverse, about 1.05-1.20 x as broad as long, less rounded at sides and less clearly contracted to base (fig. 141); segments 5 and 6 of antennal funiculus evidently transverse (fig. 142) [very common and abundant; median lobe without a subapical constriction (fig. 143)]. .... ***postica***

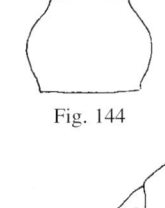

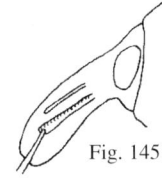

Male fore- and mid-tibia more strongly and more abruptly bent inwards at apex, inner margin strongly curved. Female fore- and mid-tibia more gradually and less strongly bent inwards at apex, inner margin less strongly curved.

Fig. 142   Fig. 143

In grasslands, waste places and wood edges and at roadsides and on cliffs. Most frequently on Medicago lupulina, *but recorded also from other Papilionaceae, particularly other* Medicago, Melilotus *and* Trifolium *species; a notorious pest of* Medicago sativa *in several parts of the world, but only causing minor damage to clover (particularly in dry years) in Britain (Gratwick, 1992). Larva pale green with a narrow, median longitudinal white line; cocoon 'latticed'. Very common and often abundant. Widely distributed throughout England, Wales and Ireland. In Scotland from the borders to Edinburgh, with an isolated record from the Outer Hebrides. Not recorded from the northern mainland, the Ebudes, Orkney or Shetland. Throughout Europe. Introduced into North America.*

13 Elytra uniformly covered with green or brown scales, often bright green to yellowish or reddish; sides, base and disc of elytra without dark brown to black marks and patches; pronotum less strongly transverse to nearly quadrate, about 1.10-1.28 x as broad as long; rostrum shorter, much shorter than pronotum [small species, 3.70-4.35 mm]. **15**

Fig. 144

- Elytra brown with conspicuous darker to black markings on disc along suture, at sides and sometimes at base; on average, pronotum more strongly transverse, about 1.16-1.30 x as broad as long; rostrum longer. **14**

14 Pronotum more strongly transverse, about 1.30 x as broad as long, and strongly rounded at sides (fig. 144); rostrum with a distinct short furrow or impression running from above antennal insertion posteriad, about as long as rostrum is wide (fig. 145); larger species, 3.9-4.8 mm [mainly on *Trifolium*]. .... ***meles***

Fig. 145

Male fore-tibia more strongly and more abruptly bent inwards at apex; mid-tibia with a stronger apical tooth; antennae inserted nearer to rostral apex, at a distance about equal to rostral width. Female fore tibia more gradually and less abruptly bent inwards at apex; mid-tibia with a weaker apical tooth; antennae inserted further from rostral apex, at a distance of about 1.5-2.0 x rostral width.

In grasslands, waste places and at roadsides. On Trifolium *spp., especially* T. pratense *and* T. repens, *but also recorded in continental Europe from* T. arvense, T. incarnatum, *etc., as well as less certainly from species of* Lotus *and* Medicago. *Larva dull green with broad yellowish median longitudinal band; cocoon 'latticed'. Formerly considered rare (RDB3), though not well- known (Fowler, 1891; Beare, 1922; Allen, 1972b). Now regarded as locally not uncommon; widely distributed in England from Dorset and S Hants. northwards to Mid-W York. Doubtfully recorded from Wales and not from Ireland or Scotland. Throughout Europe. Introduced into North America.*

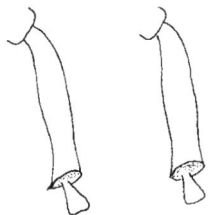

Fig. 146

Fig. 147   Fig. 148

-   Pronotum less strongly transverse, about 1.15-1.20 x as broad as long, less strongly rounded at sides (fig. 146); rostrum without, or at most with a very obscure, furrow above antennal insertion; smaller species, 3.1-3.8 mm [on *Anthyllis* and *Ulex* spp.]... **venusta**

Male fore-tibia less robust, more abruptly bent inwards at apex (fig. 147); antennae inserted nearer to apex of rostrum, at a distance roughly equal to the rostral width. Female fore-tibia more robust, less abruptly bent inwards at apex (fig. 148); antennal insertion further from apex of rostrum, at a distance equal to about 1.5 x its width.

In grasslands, open areas and heathlands, often on the coast, though not exclusively. On Anthyllis vulneraria *and species of* Ulex, *mainly* U. minor; *not generally on* U. europaeus. *Also recorded from species of* Lotus, Onobrychis, Trifolium *and* Vicia *(Fowler, 1891; Scherf, 1964; Dieckmann, 1989), but not general on these hosts in Britain. Larva yellow-green to grey-green, with a paler longitudinal median line; cocoon 'latticed'. Common and widely distributed throughout most of England and Wales; Isle of Man. In Scotland recorded from only Renfrew, E and W Sutherland and the Outer Hebrides, but probably more widespread. Widely distributed in Ireland. Europe and the Mediterranean region.*

15 On average smaller, 3.3-4.1 mm; unicolorous bright green; elytra without conspicuous brown mottling; outstanding pronotal setae shorter and more clearly semi-recumbent; upstanding pale setae of interstices shorter and mostly confined to posterior half of elytra; pronotum less strongly transverse, 1.05-1.24 x as broad as long, median longitudinal line of pronotal scales generally brighter and clearer [common and widely distributed species; on *Trifolium*]. .... **nigrirostris**

Fig. 149   Fig. 150

Male hind-tibia clearly toothed at apex (fig. 149); apical tooth of fore-tibia longer and stronger. Female hind-tibia without an apical tooth (fig. 150); apical tooth of fore-tibia shorter and weaker.

In grasslands, clover fields, waste places and roadsides. On Trifolium, *principally* T. pratense, *but recorded on the continent from other* Trifolium *spp. A minor agricultural pest, 'clover-leaf weevil' (Gratwick, 1992). Larva cream-coloured to greenish; cocoon 'latticed'. Very common and widely distributed. Recorded from most of the English and Welsh vice-counties; Isle of Man. Less well-recorded in Scotland, but occurring in most of the Islands, except Shetland and as far north as Elgin on the mainland. Widely distributed in Ireland. All Europe and the Mediterranean region. Introduced into North America.*

-   On average larger, 3.9-4.6 mm; brown, duller, sometimes with a greenish tinge; elytra often with conspicuous brown mottling or darker patches, particularly at sides and apex; outstanding pronotal setae longer and more clearly erect or semi-erect; upstanding pale setae of interstices longer and more conspicuous and not mostly restricted to posterior half of elytra; pronotum more strongly transverse, 1.14-1.29 x as broad as long, median longitudinal line of pronotal scales generally duller and less clear [scarce species, southern England and south Wales only; on *Ononis*]. .... **ononidis**

Differences between sexes as in H. nigrirostris, *though perhaps slightly less pronounced.*

A poorly understood species which has frequently been regarded as only a form of H. nigrirostris *(Morris, 1995b). Uusally on the coast in Britain, on cliffs, undercliffs and stable dunes; only very occasionally recorded inland. On species of* Ononis, *especially* O. repens. *Larva pale green, with yellowish-white midline; cocoon 'latticed'. Scarce, though sometimes abundant when found. Southern coastal counties of England and south Wales (Glamorgan and Pembroke) only, from W Cornwall to E Kent; an inland record from Surrey requires confirmation. Not recorded from Scotland or Ireland. Central and southern Europe.*

Genus *Limobius* Schoenherr

This is a genus of only three species, of which two occur in the British Isles. Their biology is very similar to that of species of *Hypera*, but the hosts are restricted to Geraniaceae. Both the British species are scarce.

Key to species

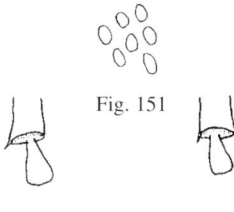

Fig. 151

Fig. 152   Fig. 153

1   Apressed scales of pronotum and elytra entire without any trace of bifurcation at apex (fig. 151); rostrum longer than pronotum; pronotum less strongly transverse, about 1.20-1.28 x as broad as long and less rounded at sides; elytra with an irregular, black, velvety patch from suture to stria 3 at about 2/5 from apex, and four irregular, short, black marks at base as well as other dark marks; size larger on average, 2.7-3.8 mm [on *Erodium*]. .... ***mixtus***

Male hind-tibia with a longer, stronger apical tooth clearly directed inwards (fig. 152); rostrum shorter, only slightly longer than pronotum, antennae inserted at about 1 x rostral width from apex. Female hind-tibia with a shorter, weaker apical tooth directed backwards (fig. 153); rostrum longer, much longer than pronotum (1.1-1.3 x), antennae inserted further, at about 1.5 x rostral width, from apex.

On stable sand dunes and vegetated fine shingle near the sea. Also recorded from the East Anglian Breckland (W Suffolk) but not recently. On Erodium cicutarium. *Larva and cocoon undescribed. Very local and restricted. S Devon, Dorset and W Suffolk (formerly, probably extinct); E Sussex and E Kent. RDB2; clearly vulnerable to coastal development. A very narrow range in the Palaearctic: recorded only from Belgium, France, Morocco, Tunisia and Algeria.*

-   Apressed scales of pronotum and elytra bifurcate at apex (fig. 154); rostrum shorter than pronotum; pronotum more strongly transverse, about 1.40-1.45 x as broad as long and more rounded at sides; elytra with scattered isodiametric black marks, irregularly disposed, and with obscure dark markings at base, but without a broad, velvety, black patch at about 2/5 from apex; size smaller on average, 2.5-3.0 mm [on *Geranium*]. .... ***borealis***

Fig. 154

Male fore-tibia bent inwards at apex, outer margin distinctly curved; rostrum shorter and thicker, antennae inserted further from base, at about 1 x rostral width from apex. Female fore-tibia not, or scarcely, bent inwards at apex, outer margin almost straight; rostrum longer and less thick, antennae inserted closer to base, at about 1.5 x rostral width from apex.

On grasslands, in rocky places and at the edges of woods and scrub. On species of Geranium, *especially* G. pratense, G. sanguineum, G. robertianum *and, probably,* G. sylvaticum; *recorded from* G. pusillum, G. molle, G. pyrenaicum, *etc., and also* Erodium cicutarium, *in central Europe. Larva yellowish-white; cocoon 'closed', yellowish. Very local, though often common where found, and widely distributed throughout England from W Cornwall to S Northumberland. Recorded from Wales (Glamorgan, Caernarvon and Flint) and with rather isolated occurrences in Scotland (Renfrew, Linlithgow, Roxburgh and Kincardine). No Irish records. Europe and the Mediterranean region to the Caucasus.*

Subfamily **Cioninae**

This is a small subfamily of five genera and about 95 species, occurring in the Old World. Representatives of two of the four genera found in continental Europe occur in the British Isles. The group has been included in Mecininae (= Gymnetrinae) by some authors, but both groups are included as tribes of Curculioninae by Alonso-Zarazaga & Lyal (1999). Here, the subfamily Cioninae is regarded as distinct following Thompson, (1992). The hosts of our species are *Verbascum* and *Scrophularia*; the ectophytic, slug-like larvae of the weevils, and their pupal cocoons, are familiar and conspicuous, particularly on *S. nodosa* and *S. auriculata*.

The Palaearctic Cionini were monographed by Wingelmüller (1937); his work was completed in 1914 but published posthumously.

Key to genera

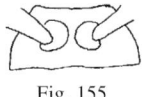

Fig. 155

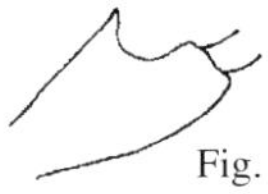

1 Elytra longer in proportion to width (about 1.3-1.5 x as long as broad; fig. 172) and flatter on disc, with an irregular, large, dark macula behind middle and scattered small black marks, but without a large, isodiametric discal spot; derm (easily seen even in unrubbed specimens) reddish-brown; anterior margin of prosternum straight (fig. 155); median tooth of fore-femur longer and sharper (fig. 156); on average smaller, 2.8-3.0 mm [male tarsal claws equal].. ***Cleopus***

Fig. 156

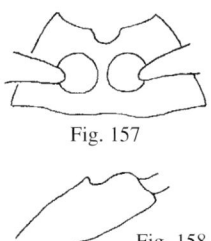

- Elytra shorter in proportion to width (about 1.0-1.05 x as long as broad; figs 160, 167, 169) and domed (convex) on disc, with a large, dark discal macula (usually circular but irregular in *alauda*) as well as a large subapical spot; derm (visible with difficulty in unrubbed specimens) black; anterior margin of prosternum deeply concave (fig. 157); median tooth of fore-femur shorter and blunter (fig. 158); on average larger, 2.8-5.0 mm [male tarsal claws often unequal]. .... ***Cionus***

Fig. 157

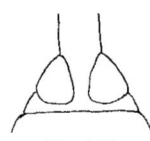

Fig. 158

Genus *Cionus* Clairville

Six species of this genus occur in the British Isles. *C. olens* (Fabricius) was included in the British list in error (Fowler, 1891) and *C. woodi* Donisthorpe, a form with deformed scales, has been synonymised with *C. scrophulariae* (Pope, 1977). Seventeen species, including several montane ones, are known from central Europe. Although the native hosts are species of *Scrophularia* and *Verbascum*, introduced Scrophulariaceae (e.g. *Phygelius*) and Buddlejaceae (mainly *Buddleja davidii* and *B. globosa*) are also foodplants.

Key to species

1 Vertex narrow, distance between eyes much less than width of rostrum at base or width of eye measured perpendicularly to rostral mid-line (fig. 159); elytra with discal black macula circular or oval, covering only the first (dilated) interstice, and without, or with less conspicuous, black marks on interstice 3 at base; predominantly blackish or grey species, if with conspicuous white pubescence this confined to sides and pronotum and to small tufts on elytra; on average larger species, 3.3-5.0 mm. **2**

Fig. 159

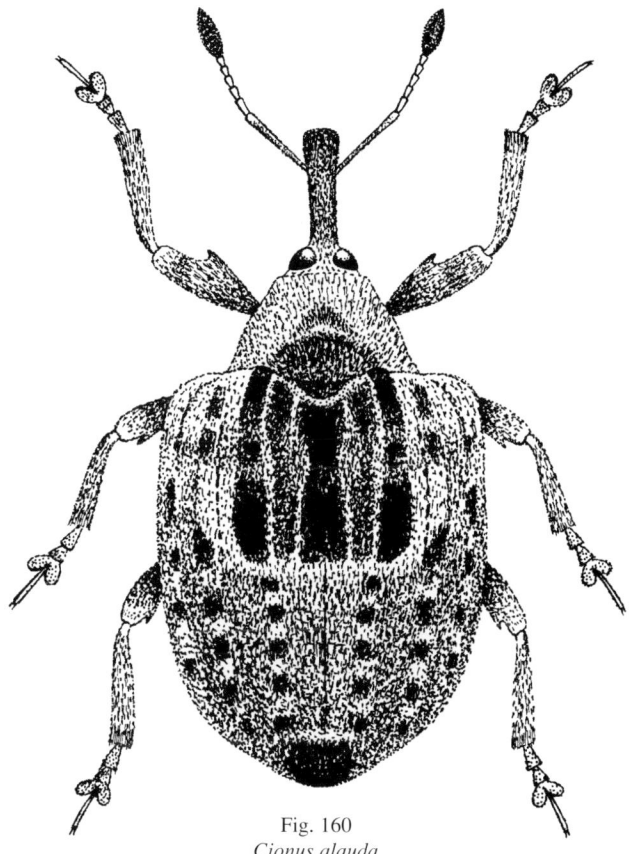

Fig. 160
Cionus alauda

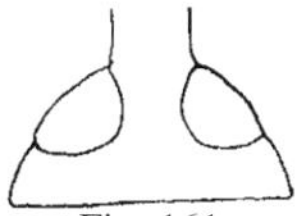

- Vertex broader, distance between eyes nearly as great as rostral width at base or of an eye measured perpendicularly to rostral mid-line (fig. 161); elytra with discal black macula irregular in shape, broader than long, covering interstices 1-3, and with a conspicuous, long dark mark on interstice 3 at base; predominantly white or whitish species, the white pubescence especially conspicuous on basal half of elytra; smaller species, 2.8-3.5 mm (habitus fig. 160). .... **alauda**

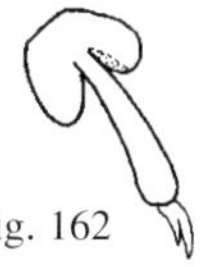

Fig. 161

Fig. 162    Fig. 163

Male tarsal claws clearly unequal, inner claw much shorter than outer (fig. 162). Female tarsal claws equal, or nearly so (fig. 163).

In and at the margins of woods, on roadsides and at the sides of ponds, lakes and watercourses. On Scrophularia nodosa *and* S. auriculata, *larvae feeding externally on the lower leaves and pupating in cocoons in the soil (Read, 1977); also occasionally on* Verbascum *spp. There is one generation a year. Locally common and widely distributed throughout England as far north as Durham; recorded from several vice-counties in south and north Wales but not from Ireland or Isle of Man. Less common in Scotland, but recorded from the Borders northwards to Fife, Main Argyll and Dunbarton. Throughout Europe and North Africa; represented in the south of its range by f.* villae *Comolli, a much more orange-coloured and less white form.*

65

2 Elytra predominantly black, light and dark tessellation less conspicuous; sutural maculae bordered with conspicuous white pubescence, discal macula with pubescence behind it, subapical macula with pubescence in front, sometimes also with a smaller patch behind. **3**

- Elytra predominantly grey to greenish-grey, light and dark tessellation more conspicuous; sutural maculae without conspicuous adjacent white pubescence. **4**

3 Pronotum thickly and uniformly clothed with white, whitish, or more rarely yellowish-white, pubescence, without a bare area on disc; thoracic epimera thickly and more uniformly clothed with pale (whitish) pubescence; elytra less strongly convex on disc; on average slightly larger, 3.5-5.0 mm (habitus fig. 167). .... ***scrophulariae***

Male rostrum shorter and duller, about as long as head and pronotum combined; distance from antennal insertion to rostral apex shorter, less than length of funiculus. Female rostrum longer and more shining, longer than head and pronotum combined; distance from antennal insertion to rostral apex longer, about equal to length of funiculus.

In woods and at their edges, at roadsides, in shady places and at the sides of ponds, streams, rivers, canals and water courses generally. Principally on Scrophularia nodosa *and* S. auriculata; *also on* S. scorodonia *and a number of introduced plants:* Phygelius capensis *(and cultivars),* Buddleja davidii *and* B. globosa. *Eggs are laid mainly in the flower buds, larvae feed on the upper leaves, and cocoons are formed chiefly among the seedheads; there is one generation a year (Read, 1977). Locally common throughout England, recorded from all the Welsh vice-counties and found in Scotland as far north as Main Argyll and Dunbarton. Not recorded from Isle of Man and occurrence in Ireland requires confirmation. Widely distributed throughout Europe and most of Asia except the far north. Introduced into North America.*

- Pronotum thickly clothed with yellow to light brownish pubescence at sides, but with a broad, almost bare, somewhat shining area on disc; thoracic epimera less thickly and more patchily clothed with brownish to pale yellowish pubescence; elytra more strongly convex on disc; on average slightly smaller, 3.4-4.2 mm. .... ***tuberculosus***

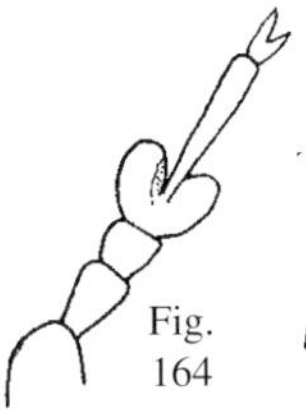

Fig. 164

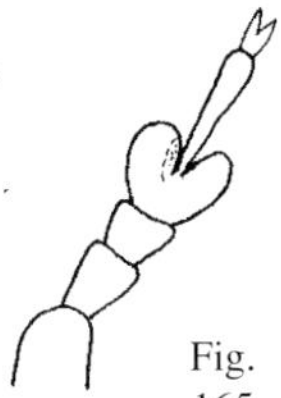

Fig. 165

Male fore-tarsi with last segment very long, as long as other tarsomeres taken together (fig. 164), claws unequal; rostrum shorter, about as long as pronotum and head combined, slightly dilated at apex, antennae inserted nearer to rostral apex (at about 2 x rostral width from its tip). Female fore-tarsi with last segment much shorter, shorter than other tarsomeres taken together (fig. 165), claws equal, or nearly so; rostrum longer, longer than pronotum and head combined, not, or very slightly, dilated at apex, antennae inserted nearer to base (at about 2.5-3.0 x rostral width from apex).

Usually in wet and marshy places and at watersides, less frequently in woods and on roadsides, etc. On Scrophularia, *particularly* S. auriculata *but also* S. nodosa; *records from* Verbascum *spp. need confirmation. Locally common in the southern counties of England northwards to E Norfolk and E Gloucester; there are also records from Leicester and Derby, and from W Perth, Main Argyll and Kintyre in Scotland. The only confirmed Welsh records are from Monmouth and Glamorgan and there are none from Ireland or Isle of Man. Throughout central and southern Europe to the Crimea.*

4 Rostrum of equal thickness throughout (fig. 166), cylindrical, longer (compare same sex), clearly longer than head and pronotum combined, strongly and more evenly curved; discal black macula usually oval, longer than broad... **5**

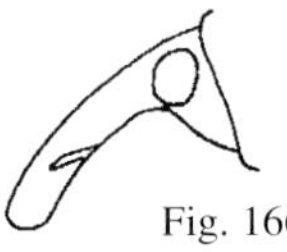

Fig. 166

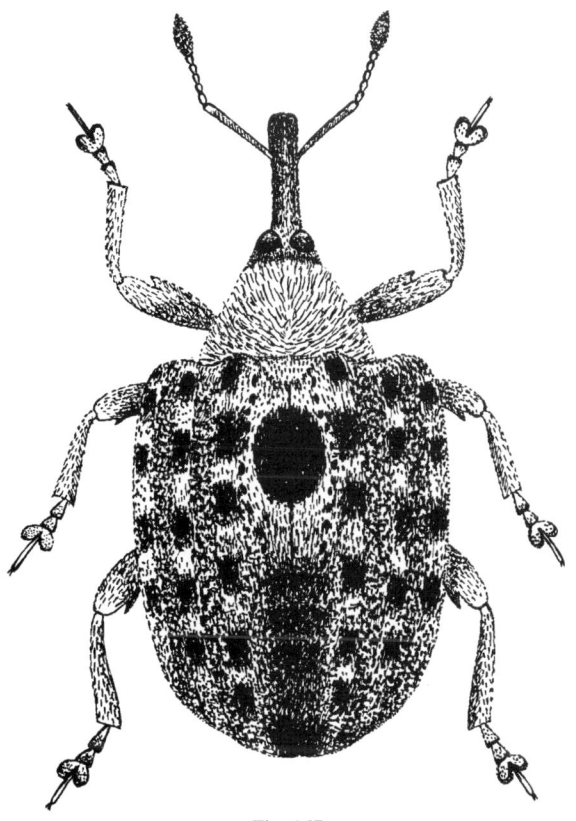

Fig. 167
Cionus scrophulariae

- Rostrum evidently subulate, especially when viewed from side (fig. 168), shorter, in male only about as long as, in female only just longer than, head and pronotum combined, more strongly and abruptly curved; discal black macula usually circular [3.8-4.6 mm] (habitus fig. 169). .... ***hortulanus***

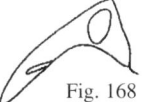

Fig. 168

Male rostrum shorter, not as long as head and pronotum combined, pubescent from antennal insertion apicad, antennae inserted well in front of middle, at about one-third from apex, apical part of rostrum from insertion shorter than scape; distal segment of fore-tarsus longer than remainder (not including claws). Female rostrum longer, as long as head and pronotum combined, glabrous and shining from antennal insertion apicad, antennae inserted just in front of middle at about 2/5 from apex, apical part of rostrum from insertion as long as, or longer than, scape; distal segment of fore-tarsus as long as, or shorter than, remainder (not including claws).

In woods and at their edges, on roadsides and in waste places. On Scrophularia nodosa *and* S. auriculata; *also on* Verbascum thapsus, *possibly other* Verbascum *and occasionally on* Buddleja *spp. There is a single generation a year. Widely distributed and often common, in Wales and southern England as far north as Nottingham and Derby. Recorded from Isle of Man and widespread in Ireland (the commonest cionine there). No Scottish records. All Europe except the far north and in Asia to western Siberia.*

67

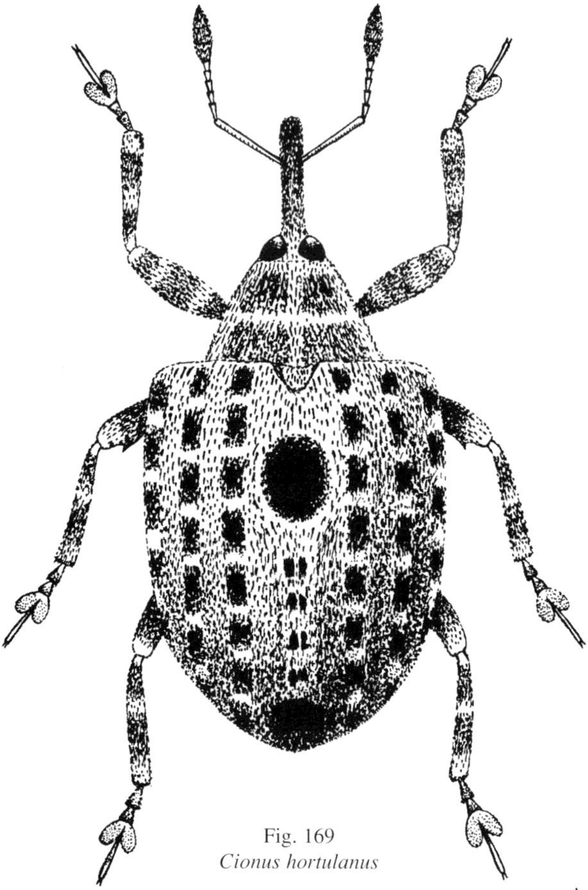

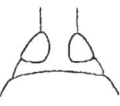

Fig. 169
Cionus hortulanus

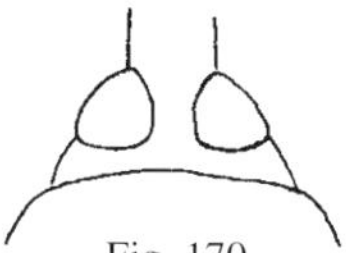

5 Subapical elytral macula larger, almost as large as discal macula; tessellation of elytral disc less distinct; eyes rounder and more prominent (fig. 170); rostrum broader and more robust, broader than distal portion of fore-femur; size larger, 4.1-4.7 mm. .... ***longicollis***

Fig. 170

Male tarsal claws evidently unequal (cf. figs 162, 164); rostrum with antennae inserted at about 1/3 from apex, apical portion shorter than scape. Female tarsal claws equal or very nearly so; rostrum with antennae inserted only a little in front of middle, apical portion longer than scape.

On roadsides and in waste places and disturbed grasslands, etc. On species of Verbascum, *especially* V. thapsus. *Local and uncommon, with two centres of distribution, occurring in Hants (N and S) and in the East Anglian Breckland (W Suffolk, W Norfolk and Cambridge). Central European and British populations belong to f.* montanus *Wingelmüller; the species is represented in south-west Europe by the nominate form (Wingelmüller, 1937).*

- Subapical elytral macula smaller, usually considerably smaller, than discal macula; tessellation of elytral disc clearer and more distinct; eyes flatter and less prominent (fig. 171); rostrum narrower and finer, narrower than distal portion of fore-femur; size smaller, 3.6-4.3 mm. ...... ***nigritarsis***

Fig. 171

68

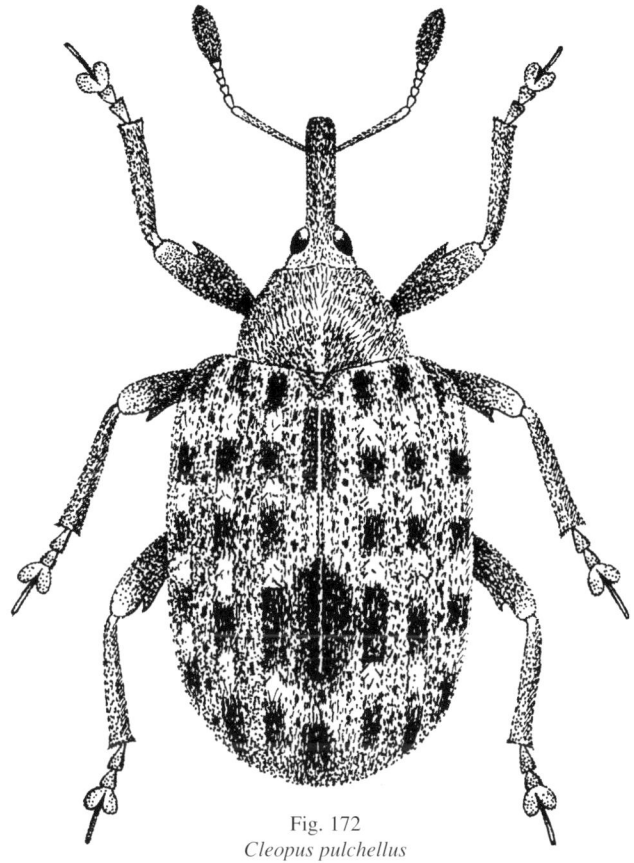

Fig. 172
Cleopus pulchellus

Male rostrum shorter, dull and pubescent almost to apex, apical part from insertion of antennae shorter than scape. Female rostrum longer, glabrous and shining to apex, apical part from insertion of antennae as long as, or longer than, scape.

In disturbed grassland in waste places, on roadsides and rough ground, etc. On species of Verbascum, *principally* V. nigrum *but also* V. thapsus. *Larvae ectophagous, mainly feeding in the inflorescences. Very local and scarce, though quite widely distributed in southern England (at least formerly) and recorded from Glamorgan in Wales. From Dorset to E Sussex northwards to Leicester, but some records require confirmation. All Europe except the far north and over much of western Asia.*

Genus *Cleopus* Dejean

Only one of the two European species of this genus inhabits the British Isles; a third species occurs in Japan.

- One species; a tessellated, black and white weevil, the pubescence often contrasting with the reddish-brown derm; 2.5-3.0 mm (habitus fig. 172)........................ ***pulchellus***

  All male tibiae armed with a fine, sharp, inwardly-pointing, apical tooth (fig. 173). Female tibiae unarmed (fig. 174).

Fig. 173

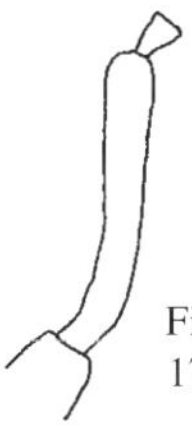

Fig. 174

69

In woods and at their margins, on roadsides, in waste and rough places and at the sides of watercourses. Most frequently on Scrophularia nodosa *and* S. auriculata, *but also on* Verbascum *spp., mainly* V. thapsus. *Larvae ectophagous, feeding mainly on the underside of leaves but also on stems, buds and flowers (Read, 1976). There are two generations a year (Read, 1976). Somewhat local, but widely distributed throughout England and Wales and extending as far north as Main Argyll in Scotland. Very local and uncommon in Ireland (S Co. Kerry, Cos Wicklow and Antrim) and not recorded from the Isle of Man. Central Europe, less common in southern Europe and N. Africa.*

Subfamily Molytinae

As recognised here, this group includes the Hylobiinae of Pope (1977), together with Pissodinae, Acicnemidinae and *Syagrius*. It excludes *Alophus*, which has had various placements in weevil classifications, and has been dealt with already (Morris, 1997). The subfamily, as redefined by Kuschel (1987) on a world-wide basis (and broadly followed by Alonso-Zarazaga & Lyal, 1999) represents the amalgamation of a large number of groups previously regarded as separate subfamilies and, perhaps as a result, has proved somewhat difficult to characterise satisfactorily. The group contains some of our largest and bulkiest species and, in these particularly, the apex of each tibia bears a strong, curved hook or tooth. Whether these processes are properly unci or mucrones is not always easy to determine, though an attempt to discriminate has been made in the subfamily key, and below, purely as a means of identification.

Because several groups formerly considered to be subfamilies have been included in Molytinae by Kuschel (1987) they are treated as tribes and subtribes here, despite most of them being monogeneric. The names of these tribes are mostly familiar but Phrynixini is not. *Syagrius* was transferred to Phrynixinae by Kuschel (1964); later (Kuschel, 1987) he included the group in Molytinae as Phrynixini.

The biological features of members of the subfamily are varied, but several species are ground-living.

Key to tribes, subtribes and genera

1   Apical tooth of tibia clearly an uncus, outer edge of tibia produced in a smooth, uninterrupted curve, inner edge of tibia clearly distinct, not forming part of tooth (figs 175, 176).. 2

-   Apical tooth of tibia a mucro, outer edge of tibia not produced in a smooth curve, either with a broad, sharp-edged rounded expansion laterally (figs 177, 178), or interrupted, in dorsal view, by setae (figs 179, 180), inner edge of tibia not distinct, forming part of tooth. 3

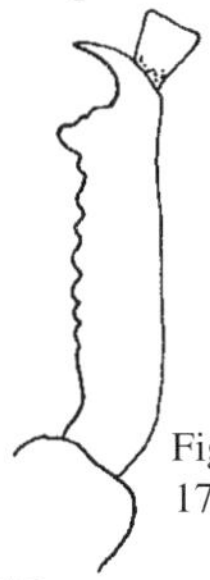

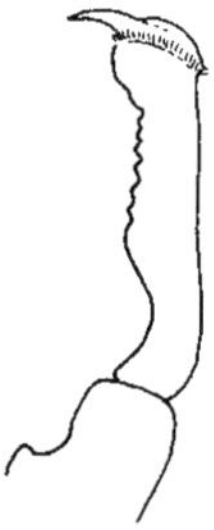

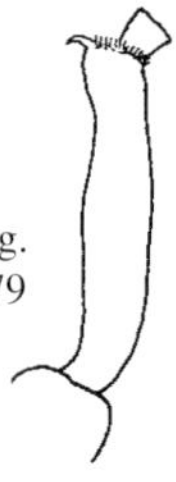

Fig. 175

Fig. 176

Fig. 177

Fig. 178

Fig. 179

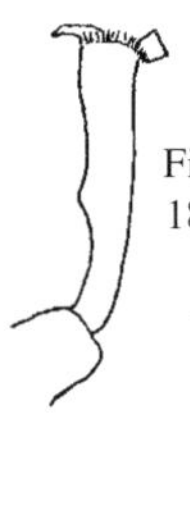

Fig. 180

2 All femora strongly toothed; elytra, pronotum and head with broad, upstanding, erect to semi-erect scales; pronotum and elytra also with broad, isodiametric, overlapping, apressed scales (fig. 181); pronotum strongly rounded at sides, its base much narrower than pronotal base, it and elytral base conspicuously depressed, best seen in lateral view (fig. 182).. **Acicnemidini**, one genus, *Trachodes*

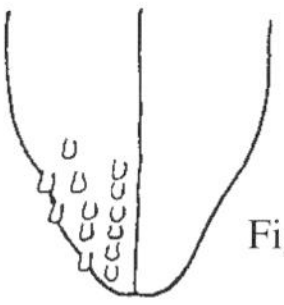

\- All femora without teeth; elytra, pronotum and head without upstanding scales; apressed scales of pronotum and elytra small, mainly elongate, if isodiametric then not overlapping; pronotum less strongly rounded at sides, its base almost as broad as elytral base, it and elytral base not conspicuously depressed seen in side view (fig. 183).
.... **Pissodini**, one genus, *Pissodes*

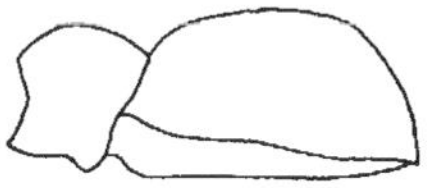

Fig. 181

Fig. 182

Fig. 183

3 Elytra with coarse nodosities, conspicuously parallel-sided, subtruncate at apex (fig. 184); pronotum tuberculate with two large nodosities on either side of disc [4.5-7.2 mm; on ferns]. .... **Phrynixini**, one genus, *Syagrius*

\- Elytra without nodosities (though with raised interstices in some cases), rounded at sides, or less conspicuously parallel-sided, apex rounded or bluntly acuminate; pronotum without tubercules or nodosities, finely to coarsely or rugosely punctured.. **4**

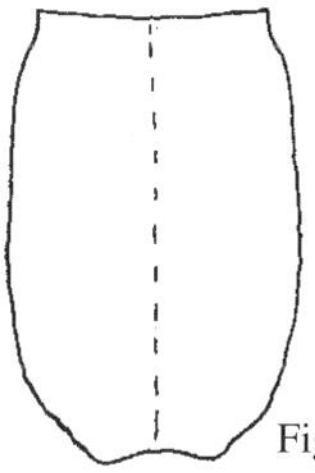

Fig. 184

4 Derm light reddish-brown; elytral interstices 2, 4 and 6 strongly raised, particularly at base and femora without teeth; pronotum strongly constricted at anterior third; [size small, 2.2 -3.0 mm; punctures of elytral striae few, very broad].
.... **Molytini** subtribe **Typoderina**, one genus, *Anchonidium*

\- Derm black to dark brown, sometimes dark red-brown; if elytral interstices raised then femora toothed; pronotum not, or scarcely, constricted anteriad. **5**

5 Pronotum with large, circular, papillate punctures, the papillae mostly off-centre and each bearing a curved, upraised, narrow seta (fig. 185); pronotal disc with a weakly raised, slightly shining longitudinal ridge, or carina, and small, shallow punctures, remainder strongly shagreened, dull; proximal segment of antennal club longer than remainder (fig. 186).
.... **Molytini** subtribe **Plinthina**, one genus, *Mitoplinthus*

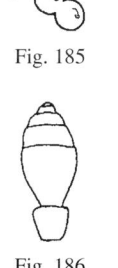

Fig. 185

Fig. 186

\- Pronotum with small, simple punctures, or rugosely punctured, without large, papillate punctures; if pronotal disc with a weak longitudinal ridge this without associated much smaller punctures, remainder of pronotum somewhat shining, not shagreened; proximal segment of antennal club shorter than remainder..................................... **6**

6 Pronotum and elytra uniformly, densely and completely clothed with narrow, apressed, variegated golden-brown to white, scale-like setae; mandibles setose; apical margin of prosternum entire; [very rare, presumed extinct species]. .... **Lepyrini**, one genus, *Lepyrus*

\- Pronotum and elytra glabrous, or with discrete patches of uniform, yellowish, narrow, apressed, scale-like setae, not completely covered; mandibles bare, or with at most one or two isolated setae; apex of prosternum emarginate. **7**

7 Elytra with distinct, well-punctured striae............................... **8**

\- Elytra without striae, entire surface reticulate or very weakly rugose with scattered, remote, small circular punctures, some bearing setae (fig. 187); [large, robust species, 12-16 mm, associated with Apiaceae] .... **Molytini**, subtribe **Molytina**, one genus, *Liparus*

Fig. 187

8 Pronotum closely, strongly and rugosely punctured; all femora strongly, but rather bluntly, toothed (fig. 188); scutellum large, conspicuous, setose; elytra with patches of yellow, recumbent, scale-like setae; larger species, 9-12 mm.. **Hylobiini**, one genus, *Hylobius*

Fig. 188

\- Pronotum simply, less strongly and closely punctured; femora with at most a small, weak, but rather sharply-pointed, tooth (fig. 189); scutellum small, inconspicuous, bare; elytra entirely glabrous, without patches of setae; much smaller species, 1.9- 3.1 mm. .... **Molytini** subtribe **Leiosomatina**,one genus, *Leiosoma*

Fig. 189

Genus *Lepyrus* Germar

This genus contains 15 species, distributed in the north of the Holarctic region. Most species appear to be polyphagous, though often associated with species of *Salix*. The Palaearctic representatives were revised by Zumpt (1936). One species has been found in Britain, but very rarely, and there are no records from the twentieth century.

\- One species: a large (8.5-10.9 mm), broad weevil; pronotum transverse, elytra rounded at sides, conspicuously narrowed and laterally impressed apicad and usually bearing an obvious patch of white pubescence on the fifth interstice about 1/4-1/3 from apex; femora bluntly toothed; other characters as in key to genera. .... *capucinus*

Underside with basal abdominal segments strongly impressed in male, impressions absent in female. In woods, wet grasslands and scrublands and steppes (in continental Europe). Polyphagous, but often associated with Salix spp. A minor pest of raspberry and strawberry in Germany. Extremely rare in the British Isles and almost certainly extinct. Nineteenth century records, almost certainly of single individuals, exist for N and S Hampshire, Surrey and Berkshire. No evidence for the putative record from Moray (Hyman & Parsons, 1992) can be traced. Widely distributed throughout northern and central Europe. Introduced into North America.

Genus *Hylobius* Germar

About 40 species are assigned to this genus, many of them associated with, and pests of, coniferous trees. Like *Lepyrus* the species are distributed throughout the northern part of the Holarctic region. The two British species contrast considerably with each other, one being an abundant, widely distributed, notorious pest of conifers ('Pine weevil'), the other a rare and endangered species associated with Purple Loosestrife, *Lythrum salicaria*.

Key to species

1 Humeral prominences and scutellum thickly clothed with conspicuous yellow, scale-like pubescence; integument red-brown; elytra with small patches of yellowish pubescence arranged in two irregular transverse bands; pronotum more deeply and coarsely, but less confluently, punctured; on average a little smaller, 7.5 -11.2 mm [rare species, on *Lythrum salicaria*]. .... ***transversovittatus***

   Male basal abdominal segments depressed, anal segment deeply excavate. Female abdominal segments not depressed.
   In fens, southern peatlands generally, undercliffs and wet places. On Lythrum salicaria, *larvae in the woody rootstocks. Very rare and endangered (RDB1). Recorded only from S Devon and N Somerset and not known to be an indigenous British species until 1944 (Ashe, 1944). No Welsh, Scottish or Irish records. Widespread in southern and central Europe, occurring as far north as Denmark; the life history in Europe was studied by Blossey (1993). A biological control agent in Canada and USA for control of its introduced host (*L. salicaria*) which is a troublesome weed there.*

- Humeral prominences without conspicuous yellow pubescence; scutellum at most with sparse, fine, white to yellowish pubescence to almost glabrous; integument black; elytra with small patches of yellowish pubescence distributed irregularly, not forming transverse bands; pronotum less coarsely and deeply, but more confluently, punctured; on average larger, 8.0-13.4 mm [common species, on conifers, especially *Pinus sylvestris*]............ ***abietis***

   Male abdominal segments depressed, anal segment deeply excavate; elytra narrower, about 1.5 x as long as broad. Female abdominal segments simple; elytra broader, about 1.4 x as long as broad.
   In coniferous woods and plantations: 'the maintenance of a high endemic population of the weevil is an inevitable result of production forestry' (Bevan, 1987). On Pinus spp., *especially* P. sylvestris *and* P. contorta, Picea *spp. and other conifers; larvae in stumps. All stages may be found throughout the year and adults may live (and breed) for two years. Adults feed on the foliage of other plants as well as pine needles. Widely distributed and abundant throughout England, Wales, Ireland and Scotland, though not recorded north of West Ross. A troublesome and intractable pest, particularly of regeneration forestry (Bevan, 1987), on which there is an extensive literature. Throughout northern and central Europe.*

Genus *Liparus* Olivier

Species of this genus are among the largest of European weevils and *L. germanus* was colloquially known as 'grandfather' to an earlier generation of British coleopterists. However, the measurements given by Fowler (1891) appear to be exaggerated. About 20 species are known, occurring mainly in mountainous regions of the Palaearctic region, but only two of these are found in the British Isles. Those species of the genus which have been examined (including the two British ones) are flightless, with reduced wings.

Key to species

1 Elytra with numerous small patches of yellowish scale-like
pubescence; most patches with about 20 setae each;
patches rather evenly distributed; thoracic disc less closely
punctured, with large and small punctures intermingled
(fig. 190); average size larger, 12.5-15.0 mm [on
Heracleum sphondylium; South-East England only].
.... ***germanus***

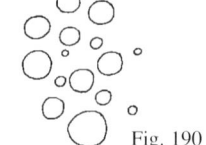

Fig. 190

*Male with first abdominal segment deeply impressed; anal segment impressed and densely clothed
with fine golden pubescence; elytra narrower, about 1.3 x as long as broad; antennae inserted slightly
closer to rostral apex. Female abdominal segments not impressed, anal segment sparsely pubescent;
elytra broader, about 1.2 x as long as broad; antennae inserted slightly further from rostral apex.*

In grasslands and hedgerows, at wood edges and in open places generally. On, or at the base of,
Heracleum sphondylium *and possibly other Apiaceae, larvae feeding in the rootstocks. Very local,
though sometimes fairly common when found. In south-east England only, recorded from E Sussex, E
and W Kent; a record from Surrey may require confirmation. Widely distributed in southern and
central Europe. RDB2.*

- Elytra often glabrous, with at most a few, inconspicuous
patches of yellowish pubescence; patches mostly with 2-8
setae; patches sparsely and irregularly distributed; thoracic
disc more closely punctured, with punctures of one size
only (fig. 191); average size smaller, 10.0-12.6 mm
[chiefly on *Anthriscus sylvestris*; widely distributed in
England]. .... ***coronatus***

Fig. 191

*Male with first and second visible abdominal segments broadly, but rather shallowly, impressed;
anal segment impressed, thickly pubescent at apex; elytra narrower, about 1.3 x as long as broad.
Female first and second visible abdominal segments less broadly and deeply impressed; anal segment
not impressed, apex not pubescent; elytra broader, about 1.18 x as long as broad.*

*In grasslands, at the sides of tracks, rural roads and woods, and in open places generally, often on
calcareous substrates. On* Anthriscus sylvestris *and possibly other Apiaceae, larvae feeding in the
rootstocks. Local and not generally common, but widely distributed throughout southern England to
Cambridge and Norfolk. There are more isolated records from Stafford, Derby and NE York. Recorded
only from Glamorgan in Wales, not from Scotland, and only doubtfully from Ireland (Morris, 1993b).
Widely distributed in central and southern Europe northwards to Denmark and southern Sweden.*

Genus *Leiosoma* Stephens

About 30 species are known in this genus from the Palaearctic region. All are small, black
weevils, in shape resembling miniature *Liparus*. They are mostly associated with
Ranunculaceae. Three species inhabit the British Isles.

Key to species

1 Femora each with a small, but sharp and distinct, tooth
(fig. 192) [common species, 2.4-3.0 mm]............................. **2**

- Femora not toothed [uncommon species, 1.8-3.1 mm]........... **3**

2 Black, tarsi and antennae, except club (black), red to
pitchy. (habitus fig. 193). .... ***deflexum*** (type form)

Fig. 192

*Secondary sexual characters not marked and somewhat variable. Male elytra slightly narrower;
rostrum a little shorter and thicker, antennae inserted a little closer to apex (distance from insertion
to apex about equal to rostral width at insertion); abdomen with basal (first) segment broadly, but*

74

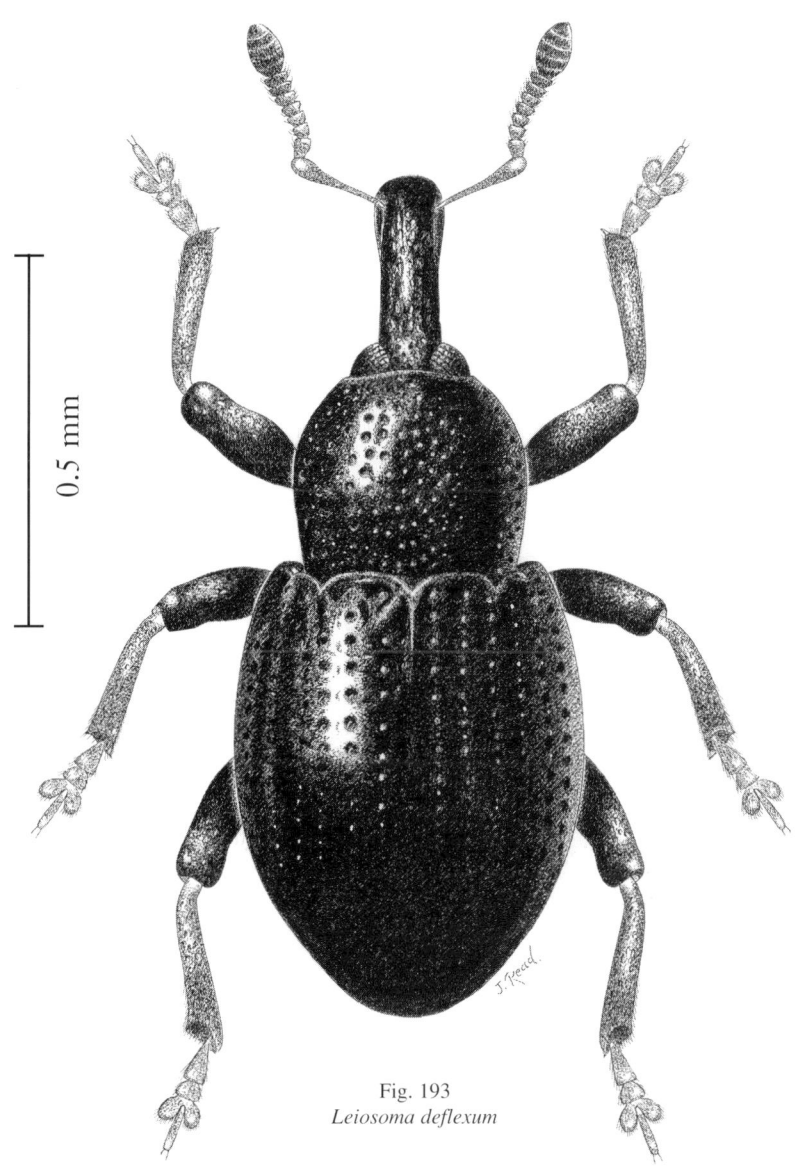

Fig. 193
Leiosoma deflexum

shallowly, impressed. Female elytra generally a little broader; rostrum a little longer and less robust, antennae inserted further from apex (distance from insertion to apex clearly greater than rostral width at insertion); basal segment of abdomen not impressed.

In woods, grasslands, hedgerows, roadsides and gardens, often in rather damp or wet areas. Ground-living, but also taken at low levels on the foodplants. On Anemone nemorosa, Ranunculus *spp. (especially* R. repens) *and* Caltha palustris, *but exact range of hosts in the British Isles not accurately known. Larvae in roots and stolons. Widely distributed and often very common throughout the British Isles, though probably under-recorded in central and northern Scotland. Known from N and S Ebudes, Hebrides and Shetland. Throughout Wales and much of Ireland. Widely distributed in central and northern Europe.*

- Red, antennal clubs, elytra, and often femoral apices, black
to pitchy. .... *deflexum* f. *collare*

Individuals of this red form are generally accepted to be merely immature examples of the typical form (but see Hoffmann, 1958, for a contrary view). The form is not mentioned by Kippenberg (1983). The size and morphological differences between deflexum *and* collare *do not appear to be constant. Other colour forms are known from France (Hoffmann, 1958). Widespread and fairly common in the British Isles.*

3 Metepisternum with a conspicuous patch of white scales;
larger (2.5-3.1 mm) and more elongate, elytra about 1.35 x
as long as broad; pronotum more shining; elytral striae less
strongly marked, punctures shallower and more distant.
.... *oblongulum*

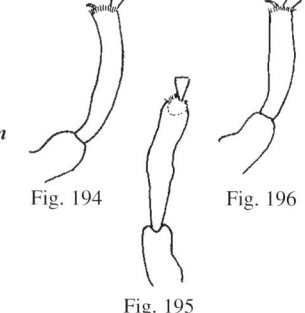

Fig. 194   Fig. 196

Fig. 195

Male fore-tibia strongly curved inwards, apical third slightly 'twisted' and more thickly pubescent (figs 194, 195); antennae inserted closer to apex of rostrum, distance from apex to insertion about equal to, or less than, rostral width. Female fore-tibia straighter, apical third not 'twisted', pubescence evenly distributed (fig. 196); antennae inserted further from rostral apex, distance to insertion greater than rostral width.

In woods, grasslands, hedgerows and on cliffs; generally in similar places to those inhabited by L. deflexum. On Anemone nemorosa *and species of* Ranunculus, *larvae probably in the roots, but hosts and larval habits little-known. Local and rather rare but widely distributed in southern England northwards to Cumberland. No records from Scotland or Isle of Man and from only Brecon and Carmarthen in Wales. Widely distributed but local in Ireland. Central and southern Europe, not recorded from Scandinavia.*

- Metepisternum without a patch of white scales; smaller
(1.8 -2.4 mm) and broader, elytra about 1.3 x as long as
broad; pronotum less strongly shining; elytral striae more
strongly marked, punctures deeper and closer. ...... *troglodytes*

Male rostrum slightly shorter and proportionately broader, antennae inserted a little nearer apex of rostrum, its apical portion little longer than wide at antennal insertion. Female rostrum slightly longer and proportionately narrower, antennae inserted a little further from apex of rostrum; its apical portion distinctly longer than wide at antennal insertion. But secondary sexual differences very slight.

In grasslands, particularly chalk downland. Probably associated with species of Ranunculus. *but foodplants and biology unknown. Very rare and scarce. Recorded only from S Hampshire, E Sussex, E Kent and Surrey. Not known from western, central or northern England, nor from Scotland or Wales. Very rare in Ireland (Co. Antrim and, less certainly, NE Co. Galway). Elsewhere the species is known only from northern France. Its status as a good species, and not a 'subspecies' of* L. pyrenaeum *Brisout (not British) was established by Tempère (1979). RDB2 (Hyman & Parsons, 1992).*

Genus *Mitoplinthus* Reitter

This genus, better known to British coleopterists as *Epipolaeus*, contains only two or three species. *Plinthus*, in which the single British species has been placed in the past, is much more speciose, with about 40 representatives in the Palaearctic region. However, most recent authorities agree in placing *caliginosus* in *Mitoplinthus*.

- One species; a large (6.1-9.0 mm), elongate weevil; elytral
striae with conspicuously large, coarse punctures; first and
second funicular segments elongate, the second as long as
the succeeding three segments (fig. 197); other characters
as in generic key. .... *caliginosus*

Fig. 197

76

Male slightly narrower, on average; abdomen with anal segment truncate behind. Female slightly broader; anal segment rounded.

In grasslands and rough places, also in woods. A pest in hop gardens and yards ('hop root weevil') (Massee, 1954). Ground-living, often occurring under stones and clods; wingless. Polyphagous, the larvae in rootstocks of herbaceous and woody plants, including Humulus lupulus, *and, less frequently, in rotten wood. Pupation normally occurs within the host, from May to August, and larvae may live two years (Collingwood, 1954). Very local, and generally rare except in the hop-growing areas of E and W Kent, Hereford and Worcester; also recorded from S Wilts., S Hampshire, E Sussex, Surrey, Middlesex and Derby. Not known from south-western, eastern or northern England, nor from Wales, Scotland or Ireland. A rather narrow range, in western-central Europe only.*

Genus *Anchonidium* Bedel

About four species are known in this genus, only one of them occurring in the British Isles. It was first taken in 1893, but not recognised until later (Keys, 1916).

- One species; a small (2.2-3.0 mm), reddish-brown species; alternate elytral interstices strongly raised, each with a single line of curved setae; prothorax, elytra, epimera and abdomen with large, circular, shallow punctures; other characters as in generic key (habitus figure as frontispiece to Hyman & Parsons, 1992). .... ***unguiculare***

Abdomen strongly impressed in male, simple in female.

In western oakwoods (i.e. on acid soil) and cliff grasslands, often found in leaf litter. Ground-living; wingless. Biology unknown. Very local, but often abundant where found. In sessile oakwoods in the Gweek area of W Cornwall and on sea cliffs in S Devon only. Not recorded from central, eastern or northern England, nor from Wales, Scotland or Ireland. A very narrow range in Europe: France and Spain only, besides England. RDB2.

Genus *Trachodes* Germar

Formerly regarded as a separate subfamily (Acicnemidinae = Trachodinae), the molytine tribe Acicnemidini contains only *Trachodes* in the Palaearctic fauna. The genus includes about 10 species, only one of which occurs in the British Isles.

- One species; a small (2.5-3.9 mm) weevil, upper surface with conspicuous broad, upstanding scales; tibiae bisinuate internally (fig. 198); pronotum with black scales on disc, sides abruptly demarcated by broad, light grey to whitish, overlapping, apressed scales; femora clavate, sharply toothed; other characters as in generic key. .... ***hispidus***

Fig. 198

Male often slightly narrower; rostrum a little shorter, slightly less strongly curved, less strongly shining and more robust; distance from antennal insertion to apex about twice length of scape. Female usually slightly broader; rostrum a little longer, more strongly curved, more strongly shining and less robust; distance from antennal insertion to apex more than twice length of scape.

In broad-leaved woodland, old hedgerows, pasture woodlands and parks. Often in faggots, especially those of Quercus *and* Corylus, *on dead wood, under bark and in leaf litter. Usually ground-living, but also taken on the trunks of trees, especially* Carpinus betulus. *Biology unknown, but likely to be associated with dead wood, particularly twigs. Uncommon and generally scarce, but widely distributed throughout England from N and S Devon and E Kent to Cumberland. Also recorded from Kirkcudbright, but not elsewhere in Scotland, and from Monmouth, Merioneth and Denbigh in Wales. Not known from East Anglia or Ireland. Widespread in Europe, including Scandinavia. Introduced into North America.*

Genus *Syagrius* Pascoe

This genus is Australian (Marshall, 1922) and, although our single species must have originated in that region, it has not been found there yet, despite having been described over 90 years ago. *Syagrius* was long included in subfamily Rhyparosominae but was removed into Phrynixinae by Kuschel (1964) (as noted previously in the introduction to Molytinae) where the genus is currently placed in tribe Phrynixini. Species of *Syagrius* feed exclusively on ferns.

- One species; a dark brown to black weevil with conspicuous nodosities on both elytra and pronotum; elytra parallel-sided (fig. 199), abruptly and obliquely angled basad, declivity markedly truncate; antennal scape very long, about equal in length to funiculus including club; length 4.5-7.2 mm (see also generic key); (colour figure in Blair, 1948).. *intrudens*

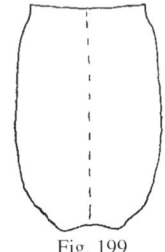

Fig. 199

Male rostrum more coarsely and closely punctured, especially basad, less strongly shining and less strongly curved, rather abruptly expanded from behind antennal insertion to apex; antennae inserted nearer apex, distance less than 1.5 x width of rostrum at insertion. Female rostrum more finely and remotely punctured, particularly at base, more strongly shining and more strongly curved, more gradually expanded from behind antennal insertion to apex; antennae inserted further from apex, distance about 2.0 x width of rostrum at insertion.

In fern-houses, ornamental parks and gardens where ferns are grown, and occasionally in open country. On a variety of introduced and native ferns, including Pteridium aquilinum *but by no means general on bracken. Larvae in stems and rootstocks; the life-history was described by Mangan (1908). Extremely local, but recorded from three or four widely separated areas of the British Isles: Ireland (Co. Dublin, where it was first recorded); W Cornwall; E Sussex (where it is persistent and not uncommon at one well-known site) and E Kent; and Glamorgan, South Wales; also recorded from Guernsey (Channel Islands) (Heijerman, 1999; Thompson & Howell, 2001), but not known elsewhere in continental Europe. RDB3 (Shirt, 1987); the proposal to delete this status (Hyman & Parsons, 1992) abrogates international responsibility for conservation of the species, which is not yet known from anywhere else in the world (see also the opinion of Heijerman, 1999).*

Genus *Pissodes* Germar

Formerly placed in a separate subfamily, the tribe Pissodini is now included in Molytinae. About 20 species of *Pissodes* are known from the Palaearctic region and about 30 from North America. They are associated with coniferous trees and several are important forestry pests. Three species inhabit the British Isles. The North American *P. strobi* (Peck) ('White pine weevil') was also included by Bevan (1987) because of its potential as a very damaging pest of conifers, particularly *Picea* spp.; it has not been found in the British Isles. Other species, especially *P. piniphilus* (Herbst), have occasionally been imported into Britain in softwood timber, but have not become established.

Key to species

1 Colour lighter, brown to brownish-orange; size smaller, 5.0 -7.2 mm; pronotum less strongly transverse, 1.10-1.15 x as broad as long, closely, but not confluently or rugosely punctured; elytra with bands of scales of two colours, white and ochreous, scales less elongate, mostly about twice as long as broad, third and fourth interstices not strongly raised at base. 2

- Colour darker, black to dark brown; size larger, 6.1-9.3 mm; pronotum more strongly transverse, 1.20-1.25 x as broad as long, rugosely and confluently punctured, especially on disc distad; elytra with bands of unicolorous scales, whitish to cream, scales more elongate, about 3 x as long as broad, third and fourth intersices strongly raised at base.. *pini*

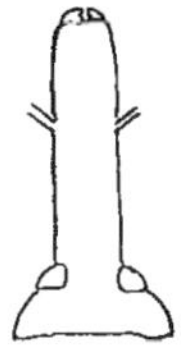

Fig. 200

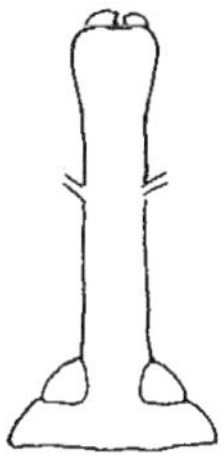

Fig. 201

Male rostrum shorter, about equal in length to pronotum, straighter and more robust and more strongly punctured, less strongly dilated apicad (fig. 200); antennal scape about equal to distance from rostral apex to insertion. Female rostrum longer, a little longer than length of pronotum, more strongly curved, more delicate, more strongly dilated apicad (fig. 201); antennal scape shorter than distance from rostral apex to insertion.

In pine woods and conifer plantations. A relatively minor pest of Pinus *spp., ('Banded pine weevil'). Particularly on* P. sylvestris, *larvae in dead stems (generally in thicker timber than* P. castaneus*); adults feed on pine foliage. Local, but often abundant where it occurs; regarded as a northern species. Throughout Scotland (though not recorded from the Islands) and northern England. More sporadic in southern England and Wales, but found in scattered vice-counties southwards to N Devon and S Hants. Not recorded from Ireland. Widely distributed in Europe.*

2   Scales of sides of prothorax, above fore-coxae, more than twice as long as broad (fig. 202), sparser, not overlapping or forming a continuous covering; base of pronotum weakly bisinuate (figs 203, 204), sides not divergent basad, without a basal tooth or slight projection; slightly smaller on average, 5.0-6.3 mm; elytral declivity steeper (fig. 205).. *validirostris*

Fig. 202

Fig. 203

Fig. 204

Fig. 205

Male rostrum slightly more robust, shorter, antennae inserted a little closer to apex. Female rostrum less robust, longer, antennae inserted further from apex.

In pine woods and conifer plantations; on Pinus *spp., especially* P. sylvestris, *but not regarded as a pine pest in Britain. Larvae feed in pine cones, not under bark of dead stems. Exclusively a Scottish species in Britain, recorded from the Highlands and Fife, though not from the Islands or southern Scotland. Local, and generally considered rare, though little-known since its discovery as a native British species (Laidlaw, 1931; Beare, 1930; Donisthorpe, 1931). Not recorded from England, Wales or Ireland. Widely distributed in Europe.*

- Scales of sides of pronotum above fore-coxae predominently isodiametric to 1.5 x as long as broad (fig. 206), less sparse, occasionally overlapping and in places forming a continuous covering; base of pronotum more strongly bisinuate (fig. 207), sides generally slightly divergent basad and prolonged into a weak and inconspicuous tooth or projection; slightly larger on average, 5.5-7.0 mm; elytral declivity less steep (fig. 208). .... *castaneus*

Fig. 206

Fig. 207

Fig. 208

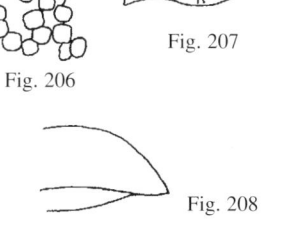

Male rostrum straighter, shorter, a little shorter than pronotum, more robust, slightly duller and more closely punctured, antennae inserted in front of middle. Female rostrum more strongly curved, longer, slightly longer than pronotum, a little less robust, slightly more shining and less closely punctured, antennae inserted at middle.

In pine woods and conifer plantations. An important pest of Pinus *spp., particularly young trees ('Small banded pine weevil'). Larvae in dead and dying wood of* P. sylvestris *and other* Pinus *spp., generally in small stems. Common and widely but patchily distributed throughout England and Wales, from W Cornwall and E Kent northwards to NE York and Durham (introduced?) and in a few highland vice-counties of Scotland. Not known from Ireland. Throughout Europe.*

Subfamily Cyclominae

The limits of this group were redefined by Alonso-Zarazaga & Lyal (1999); it was formerly known as Rhytirhininae (Thompson, 1992). Of the many variants of the spelling of the name, Rhytirrinini was authenticated by Alonso-Zarazaga & Lyal (1999); this is the cyclomine tribe to which the British species belong. The subfamily is speciose in the Mediterranean region, but *Gronops* is the only genus represented in the British Isles.

Genus *Gronops* Schoenherr

This genus includes about 20 European species, most of them found in the Mediterranean region. Two species inhabit the British Isles.

Key to species

1 Smaller, length 3.1-3.9 mm; usually paler; head only shallowly impressed between eyes, supraorbital keels weakly developed (fig. 209); pronotum less strongly, and only very slightly, transverse, about 1.05 x as broad as long, only weakly expanded distad; interstices 3 and 5 of elytra more clearly raised into uniform keels; elytra with two transverse bands of light-coloured scales, one just in front of middle the other subapical, but both variable and frequently obscured by encrustation [on Caryophyllaceae] .... *lunatus*

Fig. 209

Abdomen impressed at base in male, simple in female.
In saltmarshes, at their edges and on sand flats etc. Also found inland in arenaceous areas. On Caryophyllaceae, particularly Spergularia media *and* S. marina; *also recorded from* Spergula arvensis *and* Cerastium *spp., but hosts recognised only recently in Britain. Larval feeding habits unknown. Local, but often abundant when found. In England and Wales from N and S Devon and E Kent northwards to S Lancaster and SE York. No records from Isle of Man, Scotland or Ireland. Throughout most of Europe except the far north.*

- Larger, length 3.9-4.7 mm; usually darker; head deeply impressed between eyes, supraorbital keels sharp, prominent and strongly developed (fig. 210); pronotum more strongly transverse, about 1.3 x as broad as long, more strongly dilated distad; interstices 3 and 5 of elytra with nodosities but less clearly raised into keels; elytra black with patches of pale scales which form much less definite transverse bands in some specimens but are less frequently obscured by incrustation [on Chenopodiaceae]. .... *inaequalis*

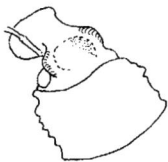

Fig. 210

Abdomen slightly impressed at base in male, simple in female.
In waste places and ruderal situations. On Chenopodiaceae, particularly Atriplex prostrata. *Almost certainly accidentally introduced but established at one site in E Kent where it is not uncommon. First found 1982 (Clemons, 1983). Also recorded from the coasts of Suffolk and N Lincoln. A species originating in eastern Siberia which has spread westwards with great rapidity during the past 40 years and is now widely, if sporadically, distributed throughout much of central and western Europe.*

Subfamily Magdalidinae
(Mesoptiliinae tribe Magdalidini)

This is a well-defined group, with some 14 genera and about 250 species which occur in all zoogeographic regions except the Afrotropical (Thompson, 1992). It is included in a subfamily Pissodinae by some authors (e.g. Lohse, 1983), but that group is currently subsumed (as tribe Pisssodini) in Molytinae (Kuschel, 1987). Alonso Zarazaga & Lyal (1999) include the group (as a new placement) as a tribe of Mesoptiliinae, a group unfamiliar in the British literature. The tribe Magdalidini is represented in Britain and continental Europe by one genus, *Magdalis*.

Genus *Magdalis* Germar

Over 100 species of this genus are known world-wide, most being Palaearctic in their distribution. All the species are associated with trees and shrubs, the larvae feeding under the bark of dead or dying twigs and branches. Unlike many other xylophagous beetles the adults are found most frequently on the foliage of the hosts. Most have a short season of adult life, which is late May to the end of June in southern England, although stragglers may be found later; in Scotland and northern England the times of appearance are later.

Eight species occur in the British Isles, one being a recent colonist or introduction. *M.* (*s. str.*) *violacea* (Linnaeus) has been erroneously recorded from Britain (Fowler, 1891; Allen, 1982).

Subgenera of *Magdalis* are well-defined and, having been used in much of the continental European literature, are included here (following Barrios, 1986), despite our fauna being a small one. The most speciose subgenus in the British Isles and continental Europe is *Magdalis s. str.* Species in this subgenus are associated exclusively with coniferous trees, but are apparently of no significance as forestry pests, none of our species being included in Bevan (1987). Those in other subgenera feed on various broad-leaved trees and shrubs. Species of *Magdalis s. str.* exhibit little sexual dimorphism, whereas this is marked in most of the other subgenera, especially *Edo* and *Panus*.

In common with many other xylophagous Coleoptera specimens of *Magdalis* vary greatly in size with giant as well as dwarf individuals occurring not infrequently. Consequently, although indications of size differences between species may be helpful in their identification, they cannot be regarded as absolute.

Key to subgenera and species

Fig. 211

Fig. 212   Fig. 213

1   Fore-femora strongly and conspicuously toothed (fig. 211); average size larger, 3.0-8.2 mm. **2**

-    Fore-femora without a tooth (fig. 212), or with at most a small, weak, inconspicuous one (fig. 213); average size smaller, 2.4-4.2 mm. **6**

Fig. 214   Fig. 215

Fig. 216   Fig. 217

2   Pronotum with one or more tubercles, nodosities or asperities at the subapical lateral margin (figs 214, 222), and with anterior margin little narrower than base, sides not convergent anteriad; on *Ulmus* and *Betula*.
    .... (*Odontomagdalis*) **3**

-    Pronotum without tubercules, anterior margin evidently narrower than base, sides convergent anteriad (figs 215-217); on *Pinus*, predominantly *P. sylvestris*.
    .... (*Magdalis s. str.*) **4**

3 Elytral interstices flat or weakly convex, much wider than striae, transversely striate, less strongly shining, strial punctures smaller and shallower; pronotum with a sharp, long, lateral tubercle in the apical quarter, clearly evident from above, interrupting the pronotal outline in dorsal view, surrounded by a group of smaller asperities (fig. 222); sides of pronotum less strongly rounded rather abruptly contracted to base; smaller species, 2.8-5.1 mm; on *Ulmus* spp. .... ***armigera***

Male rostrum short, about as long as head and much shorter than pronotum, dilated in middle; rostrum with antennae inserted at about 1/3 to 2/5 from apex, apical part shorter than antennal scape. Female rostrum longer, longer than head and only slightly shorter than pronotum, cylindrical, not dilated in middle; rostrum with antennae inserted at or near middle, apical part longer than antennal scape.

In woods and at their margins and particularly in hedgerows. On Ulmus *spp., particularly* U. procera; *larvae in galleries under bark of twigs and branches. Adults are occasionally found on apple trees (Massee, 1954; Alford, 1984). Common and widespread throughout England, but with few records from Wales (Monmouth, Glamorgan, Radnor and Cardigan, or Ireland (S Co. Kerry, Cos. Waterford, Offaly and Dublin), none from Isle of Man, and in Scotland reported only from Lanark and, more doubtfully, W Perth (Murray, 1853). All Europe to the eastern Palaearctic.*

- Elytral interstices strongly convex, about as wide as striae, less strongly cross-striate and so more strongly shining, strial punctures large and deep; lateral tubercle of pronotum smaller, less distinct from associated asperities and not, or scarcely, interrupting the pronotal outline seen from above (fig. 218); sides of pronotum more strongly and more evenly rounded (fig. 218), less abruptly contracted basad; larger species 3.1-6.2 mm; on *Betula* spp.. ***carbonaria***

Fig. 218

Male rostrum dull, short, a little shorter than pronotum, contracted in middle and thickened at antennal insertion, which is about 1/3 rostral length from apex, apical part of rostrum much shorter than antennal scape. Female rostrum somewhat shining, longer than pronotum, not, or scarcely, contracted in middle, cylindrical (slightly expanded at apex), antennae inserted a little on front of middle, apical part of rostrum longer than antennal scape.

In woods, moorlands where the hosts grow, and hedgerows. On spp. of Betula, *boring into dead logs, large branches and trunks. Local, though often abundant where found; generally regarded as a species of northern England and Scotland (extending as far north as Moray and Easterness, but also occurring in many southern English counties from E Kent westwards to Dorset and S Somerset as well as the Midlands. Not recorded from Isle of Man and known only from Radnor, Cardigan, Carmarthen and Caernarvon in Wales and Co. Wicklow in Ireland. Throughout the Palaearctic region.*

4 At least elytra bluish or purplish; each elytron weakly rounded at base, with a very shallow subbasal depression, and not, or scarcely, overlapping pronotum (figs 219, 220); pronotal disc less strongly and more closely punctured; punctures of elytral striae small and shallow, much narrower than interstices; smaller species, 3.0-6.1 mm; [confined to Scotland and northern England, not recorded south of Mid-W York]............... **5**

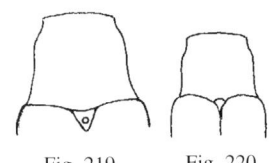

Fig. 219   Fig. 220

- Elytra black, without any trace of bluish or purplish coloration or reflection; each elytron strongly rounded at base, with a deep subbasal depression, and clearly overlapping pronotal base (fig. 221); pronotal disc more strongly and less closely punctured; punctures of elytral striae large and deep, about as broad as interstices; larger species, 4.5-8.2 mm; [a recent discovery, recorded only from southern England (E and W Sussex, N Hampshire and Surrey)]. .... ***memnonia***

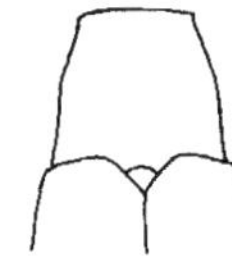

Fig. 221

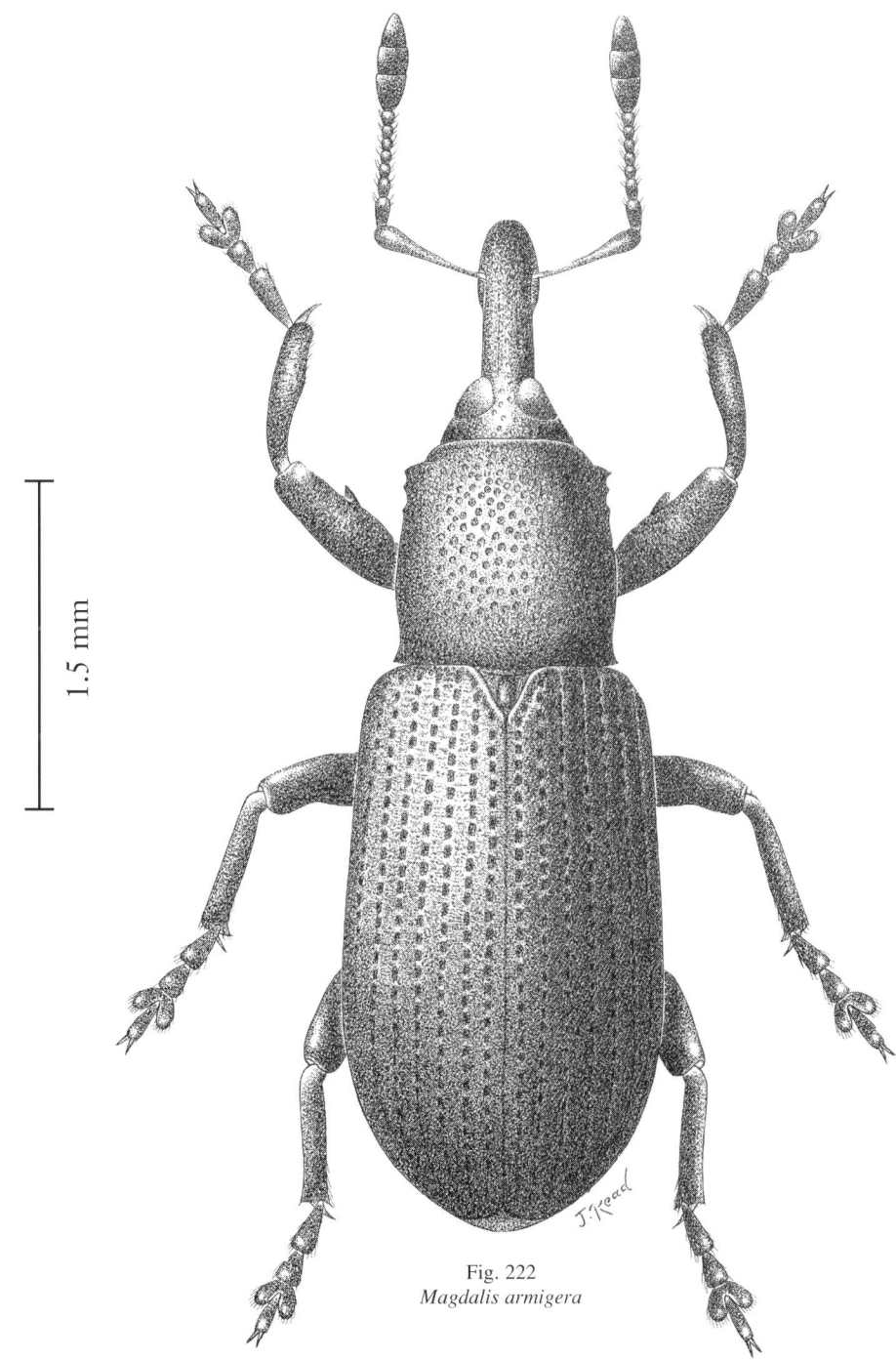

Fig. 222
Magdalis armigera

Male rostrum shorter, slightly shorter than pronotum, apical portion from antennal insertion only a little longer than antennal scape. Female rostrum longer, about as long as pronotum, apical portion from antennal insertion about 1.25 x as long as antennal scape.

In pine plantations and coniferous woodland. On Pinus sylvestris *and possibly other* Pinus *spp. Very local, in southern England only, first recorded from Friston Forest, E. Sussex (Allen, 1972a; 1976), and now known from a few other sites in E and W Sussex, N Hampshire and Surrey. A recent colonist (first record 1971), possibly introduced in forestry operations. Central and southern Europe to Siberia.*

5   Eyes flat or very weakly rounded; pronotum transverse (fig. 223); fore-femora not constricted at base (fig. 224); [3.0-4.7 mm].. ***duplicata***

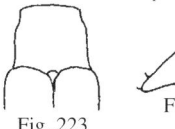
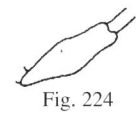

Fig. 223   Fig. 224

Male rostrum shorter and straighter, about as long as pronotum; rostrum with antennae inserted a little in front of middle, apical portion about as long as scape. Female rostrum longer and more strongly curved, slightly longer than pronotum, rostrum with antennae inserted a little behind middle, apical portion longer than scape.

In old pine woods, mainly the Caledonian Forest remnants, less frequently in coniferous plantations. On Pinus sylvestris, *the larvae in passages under the bark, penetrating into the sapwood. Local in Britain; recorded from Cumberland and Mid-W York, in northern England and a block of Scottish vice-counties in the eastern and central Highlands from Fife and Stirling northwards to Easterness and Westerness. No records from Wales, Ireland or Isle of Man. Throughout the Palaearctic region from western Europe to Japan.*

-    Eyes moderately rounded; pronotum elongate (fig. 225); fore-femora constricted at about 1/4 from base (fig. 226); [3.1-6.1 mm].. ***phlegmatica***

Male rostrum shorter and relatively more robust, not dilated apicad, about as long as pronotum, apical portion from antennal insertion a little shorter than scape, antennae inserted at middle. Female rostrum longer and relatively more delicate, gradually dilated to apex, slightly longer than pronotum, apical portion from antennal insertion appreciably longer than scape, antennae inserted behind middle.

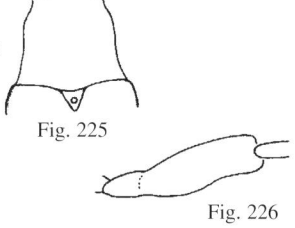

Fig. 225

Fig. 226

In old pine woods, especially the Caledonian Forest remnants, as for M. duplicata. *On* Pinus sylvestris. *British distribution similar to that of* M. duplicata; *Cumberland, S Northumberland and Mid-W York in northern England and from W Lothian and Lanark northwards to Easterness and Westerness. Not reported from Wales, Ireland or Isle of Man. Throughout most of the Palaearctic region from Europe to Mongolia.*

6   Sides of pronotum simple; rostrum weakly but distinctly curved, longer than head, or if about as long then antennal club abnormally long (males, figs 227, 228).......................... **7**

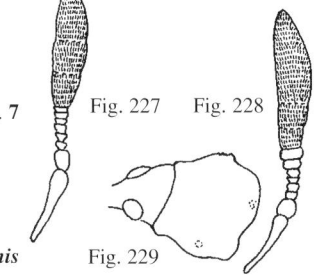

Fig. 227   Fig. 228

-    Sides of pronotum with a sharp granulate tooth just behind middle (fig. 229); rostrum straight, short, shorter than head in both sexes; [antennal club shorter than either remainder of funiculus or scape] [2.4-3.8 mm].............. (***Edo***) ***ruficornis***

Fig. 229

Male rostrum slightly shorter, distinctly contracted behind middle, dorsal surface flat with a weak depression or shallow but broad stria between antennal insertions, dull throughout, antennae inserted a little in front of middle. Female rostrum a little longer, cylindrical, not, or less strongly, contracted behind middle, dorsal surface convex, without a depression between antennal insertions, at least apical half somewhat shining, antennae inserted a little behind middle.

In woods, hedgerows and neglected orchards. On rosaceous trees and shrubs, particularly Crataegus *and* Prunus *spp. (including* P. domestica*), larvae in dead twigs. A very minor pest of plum trees and other top fruits ('Plum weevil') (Alford, 1984). Rather local but fairly common throughout England and Wales. Very uncommon in Scotland, old records only, from W Lothian and Stirling. Not reported from Ireland or Isle of Man. Europe to Siberia and Mongolia.*

7 Antennae black; tarsal claws toothed (fig. 230); all femora with a small, weak tooth, seldom obsolete (fig. 231); pronotum more strongly transverse and more rounded at sides; [basal two segments of male antennal club, taken together, longer than remainder] [2.4-4.2 mm].
.... (*Porrothus*) *cerasi*

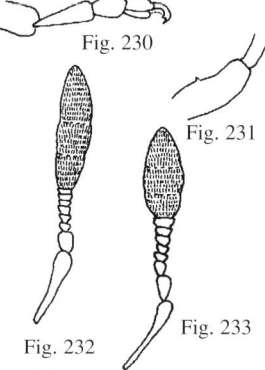

Fig. 230

Fig. 231

Male antennal club longer than either remainder of flagellum or scape (fig. 232); rostrum shorter, about as long as, or shorter than, fore-tibia, apical portion about as long as scape, constricted behind antennal insertion; elytra parallel-sided. Female antennal club normal, shorter than remainder of flagellum and about as long as, or shorter than, scape (fig. 233); rostrum longer, cylindrical, longer than fore-tibia, apical portion longer than scape, not constricted behind antennal insertion; sides of elytra divergent posteriad.

Fig. 232   Fig. 233

In woods and at their margins, in hedgerows and scrub. On species of Quercus and, perhaps less frequently, rosaceous trees and shrubs. The association of adults with oaks appears to be exclusive to Britain; elsewhere the weevil has been recorded only from Rosaceae: Crataegus, Malus, Pyrus, Mespilus, Prunus and Sorbus spp. It is not an important pest of fruit trees in Britain (Alford, 1984). Larvae in twigs and small branches. Local and uncommon, but widely distributed in England from N Somerset eastwards and northwards to NE Yorks. Not recorded from the South-west peninsula, Wales, Scotland, Ireland or Isle of Man. Throughout most of the Palaearctic region.

- Antennae entirely red; tarsal claws simple (fig. 234); all femora unarmed; pronotum less strongly transverse and less strongly rounded at sides; [basal two segments of club of male antenna shorter than remainder (fig. 235)] [3.1-4.1 mm] .... (*Panus*) *barbicornis*

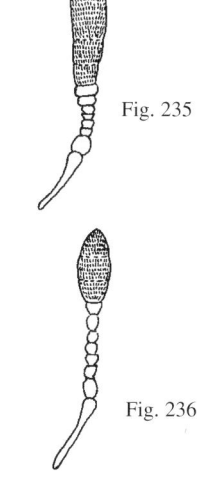

Fig. 234

Fig. 235

Fig. 236

Male antenna abnormal (fig. 235), club very long, longer than either remainder of flagellum or scape and little shorter than both taken together; two apical segments of funiculus (segments 5 and 6) strongly transverse and markedly broader than segments 2-4, first funicular segment markedly globular; rostrum short, shorter than head, dull, and constricted behind antennal insertion; pronotum distinctly less transverse, about 1.0 : 0.95; elytra parallel-sided and more elongate, about 1.8 x as long as broad. Female antenna normal (fig. 236), club shorter than either remainder of flagellum or scape, two apical segments of funiculus less strongly transverse, not, or little, broader than segments 2-4, first funicular segment not, or less clearly, globular; rostrum longer, longer than broad, somewhat shining and cylindrical, not, or scarcely constricted behind antennal insertion; pronotum more strongly transverse, about 1 : 0.85; elytra with sides divergent posteriad, less elongate, about 1.5 x as long as broad.

In woodlands and at their edges, hedgerows, scrub and neglected orchards. On a wide range of rosaceous trees and shrubs, including Malus, Pyrus, Prunus, Sorbus, Mespilus and Crataegus spp. A very minor pest of cultivated Pyrus and other fruit trees ('Pear weevil') (Massee, 1954; Alford, 1984). Larvae in subcortical chambers, pupating in spring (Alford, 1984). Generally scarce and local, but quite widely distributed in southern England from N Somerset eastwards and northwards to N Lincoln. Not recorded from the South-west peninsula, northern England, Wales, Scotland, Ireland or Isle of Man. Throughout Europe to the Crimea; Madeira, Algeria. Introduced into North America.

Subfamily Anoplinae

The absence of tarsal claws (fig. 237) characterises this small subfamily, which contains only the single genus *Anoplus*. The affinities of the subfamily are uncertain. It is sometimes placed near Ramphinae (= Rhynchaeninae), both groups having leaf-mining larvae, and sometimes in Curculioninae (*s. l.*); Alonso-Zarazaga & Lyal (1999) adopted the latter course.

Genus *Anoplus* Germar

Only three species have been described in this genus. They are small weevils, entirely black (except for the antennal scapes) and associated with Betulaceae. Two species inhabit the British Isles; *A. setulosus* Kirsch has not been found in north-west Europe.

Key to species

1   Smaller, length 1.7-2.2 mm; pronotum not shagreened between punctures, shining; setae of elytral interstices fine, long, tapering and darker (on *Betula*; habitus fig. 238). ....*plantaris*

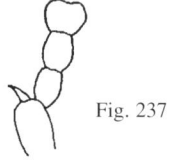

Fig. 237

Secondary sexual differences not marked. Antennae inserted slightly nearer apex of rostrum in male compared with female.

In woods, on heathlands, fens and bogs where the foodplants grow. On Betula *spp., probably equally abundant on* B. pendula *and* B. pubescens *and their hybrids. Larvae in linear leaf-mines, the mine deforming the leaf by inhibiting development of the tip. Generally common and widely distributed throughout the British Isles northwards to W Sutherland, but not recorded from Isle of Man, Hebrides, Orkney or Shetland. Widespread and common in Ireland. Throughout Europe.*

-   Larger, length 2.3-2.8 mm; pronotum strongly shagreened between punctures, dull; setae of elytral interspaces coarser, shorter, truncate at apex and paler [on *Alnus*]. .... *roboris*

Male rostrum shorter and straighter, about as long as fore-tibia; antennae inserted closer to rostral apex, at a distance about equal to breadth of rostrum at insertion. Female rostrum longer and more curved; antennae inserted further from rostral apex, at a distance greater than rostral breadth at insertion.

In wet woodland, at stream sides, and in bogs and fens. Often in much wetter places than A. plantaris. *On* Alnus glutinosa, *larvae in linear leaf-mines. Local, and generally considered rare, but widely distributed in the British Isles. In England and Wales from S Hants eastwards to W Kent and northwards to Cumberland. Local in Scotland from Dumfries northwards to Easterness, but not known from any of the Scottish Islands or Isle of Man. Local in Ireland (Cos Kilkenny, Fermanagh, E Mayo and Derry). Throughout Europe.*

Subfamily Cossoninae

This group is well-defined and generally distinctive. It is a cosmopolitan subfamily of some 1700 known species distributed among about 300 genera (Thompson, 1989). The group is retained as a subfamily by Alonso-Zarazaga & Lyal (1999). Cossonines are well represented in many island faunas and most are inhabitants of the dead wood of trees, although some occur in the woody parts of herbaceous plants or on roots. Many species appear to be dispersed in floating wood and determining the origins of some groups can be difficult. Perhaps partly because of these features the British fauna includes several species which have been introduced or which are thought to be recent colonists. The status of one of these, *Macrorhyncolus littoralis*, a recent discovery in Britain (Welch, 1990), may not

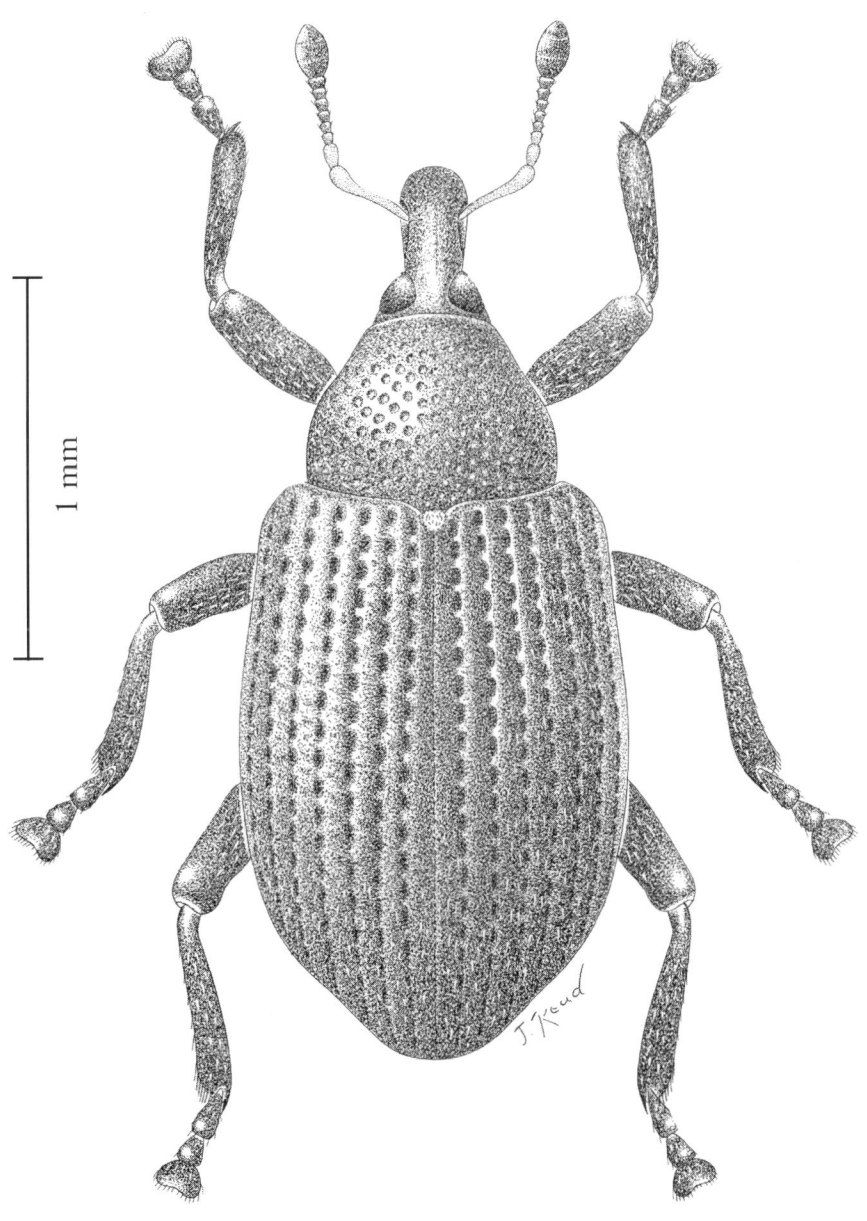

Fig. 238
Anoplus plantaris

yet be determinable. On the other hand, *Pentarthrum elumbe* (Boheman) (= *cylindricum* Wollaston), which was once thought to be a likely colonist (Buck, 1948), has not become established and is not treated fully in this account. *Caulophilus oryzae* could have been omitted also, but is included as it has been retained in British checklists such as Pope (1977).

The generic assignment of species of Cossoninae has been, and continues to be, controversial. It is, therefore, difficult to be precise about the numbers of described species currently placed in the different genera. Some nomenclatural changes have resulted from re-assessment of the groups described in Wollaston's monograph (1873) and the arrangement of Alonso-Zarazaga & Lyal (1999) shows considerable differences from previous classifications.

Key to tribes and genera

1 Antennal funiculus with five segments (fig. 239); pronotum broader than elytra [pronotum strongly rounded at sides and broadest behind middle; small species, 2.5-4.0 mm]. .... **Pentarthrini 2**

- Antennal funiculus with seven segments (fig. 240); pronotum narrower than elytra. **3**

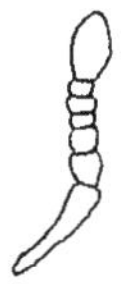

Fig. 239

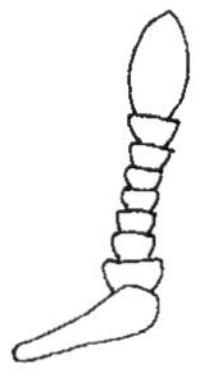

Fig. 240

2 Ninth elytral interstice strongly raised at apex, forming a long flange or keel, so sides of elytra apicad apparently explanate (fig. 241) [smaller species, 2.5-3.6 mm]. .... *Euophryum*

- Ninth elytral interstice simple, elytra not explanate apicad [larger species, 2.9-4.0 mm] .... *Pentarthrum*

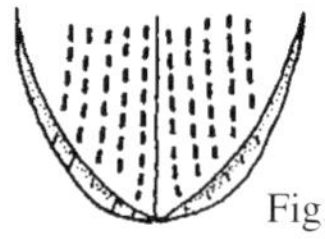

Fig. 241

3 Pronotum and elytra glabrous, or at most with a few sparse, short setae; elytra parallel-sided, or nearly so, not contracted to base. **4**

- Pronotum and elytra with copious, close, fine, pale, curved setae; elytra rounded at sides and contracted to base (fig. 242) [rostrum cylindrical; scutellum invisible; 2.6-3.8 mm]...................................... **Onycholipini** (in part), *Pselactus*

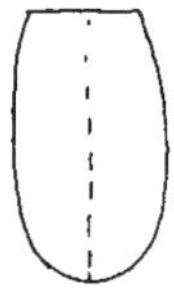

Fig. 242 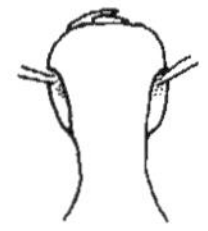   Fig. 243

4 Rostrum strongly dilated, either at antennal insertion or subapically (figs 243-245, 273); larger species (4.3-12.0 mm); elytra depressed on disc. .... **Cossonini 5**

- Rostrum without dilation, cylindrical, often short (figs 246-250, 277); smaller species (2.5-4.2 mm); elytra somewhat convex on disc. **6**

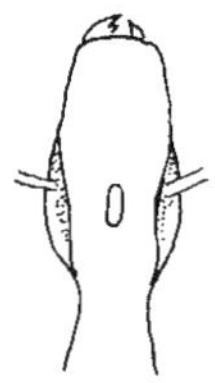

Fig. 244 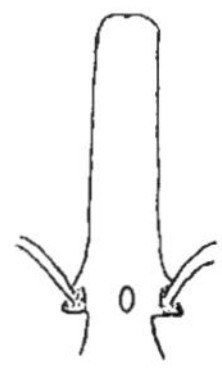   Fig. 245

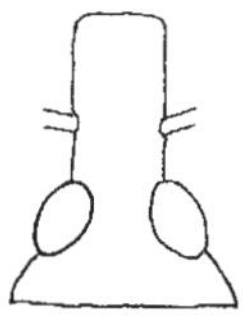

Fig. 246    Fig. 247    Fig. 248 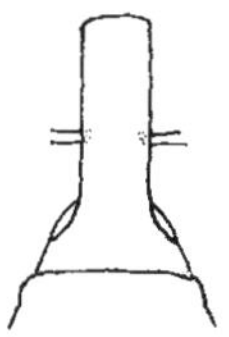   Fig. 249 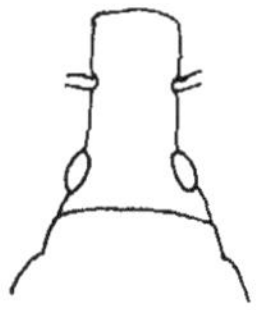   Fig. 250

88

5   Rostrum apicad broad, square, flat and spade-shaped (fig. 251); smaller species (4.3-6.6 mm); pronotum and elytra more evidently depressed.. ***Cossonus***

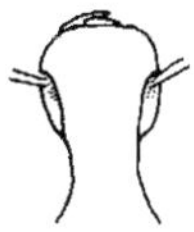

  -    Rostrum either wedge-shaped, chisel-like and dull (male, figs 252, 273); or simple, cylindrical, narrow and shining (female, fig. 253); larger species (4.3-12.0 mm); pronotum and elytra less strongly depressed. .... ***Rhopalomesites***

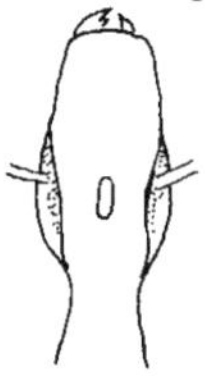

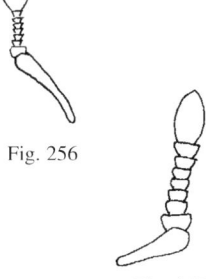

Fig. 251

Fig. 252 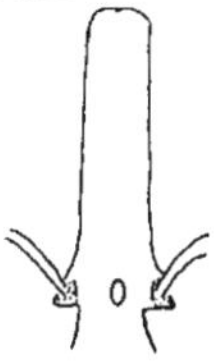   Fig. 253

6   Antennae robust, black to pitchy, club indistinct, not markedly broader than apex of scape, apical funicular segments strongly transverse (fig. 254); scutellum small but distinct; less strongly shining, without a bronzy reflection. 7

  -    Antennae fine and delicate, reddish-yellow, club distinct, pyriform, twice (or more) as broad as scape at apex, apical funicular segments quadrate to slightly transverse, globular (fig. 255); scutellum invisible; more strongly shining, with a slight bronzy reflection [2.9-3.8 mm].
       .... **Onycholipini** (in part) ***Pseudophloeophagus***

Fig. 254

Fig. 255

7   Pronotum more elongate, 1.15-1.35 x as long as broad, less strongly rounded at sides, almost straight, base little broader than apex; rostrum less elongate, 1.1-1.3 x as long as broad, broader, about half the width of pronotum.
       .... **Rhyncolini-Rhyncolina 8**

  -    Pronotum less elongate, 1.05-1.10 x as long as broad, more strongly rounded at sides, base clearly broader than apex; rostrum more elongate (except male *S. truncorum*), 1.7-2.1 x as long as broad, narrower, much less than half width of pronotum. **9**

8   Smaller (2.8-3.1 mm) and more delicate; antennal club much broader than funicular segments (fig. 256); sides of pronotum more clearly convergent to apex [very rare introduced species, southern England]. .... ***Macrorhyncolus***

Fig. 256

  -    Larger (3.2-4.2 mm) and more robust; antennal club little broader than funicular segments (fig. 257), indistinct; sides of pronotum more rounded, less clearly convergent to apex [not uncommon, mainly northern species, usually in dead *Pinus*] .... ***Rhyncolus***

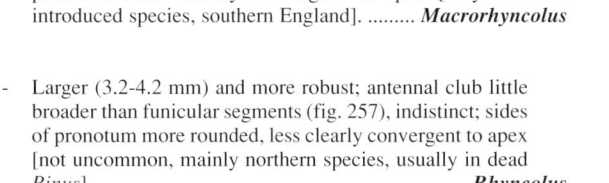

Fig. 257

9 Eyes large and prominent, dorso-lateral, distance between them little more than width of eye (as seen from above; fig. 258); pronotum more strongly contracted apicad [occurring only under artificial conditions].
.... **Dryotribini**, one genus *Caulophilus*

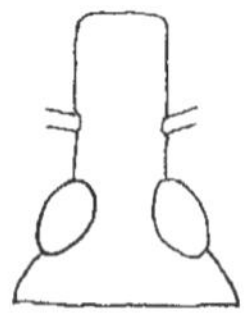

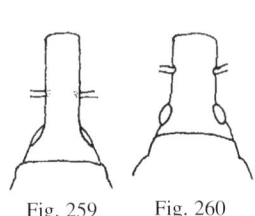

Fig. 258

- Eyes smaller and less prominent (figs 259, 260, 277), lateral, distance between them much greater than width of eye (eye sometimes invisible from above); pronotum less strongly contracted apicad [occurring outdoors, under natural conditions]. **10**

Fig. 259   Fig. 260

10 Antennal club truncate at apex, pubescent only at apex (fig. 261), mainly glabrous; eyes flat, depressed, not visible from above......... **Onycholipini** (in part) *Stereocorynes*

- Antennal club bluntly pointed at apex, pubescent throughout (fig. 262); eyes distinctly convex, or if somewhat flat then visible from above (figs 259, 260, 277)
· .... (**Rhyncolini-Phloeophagina**) *Phloeophagus*

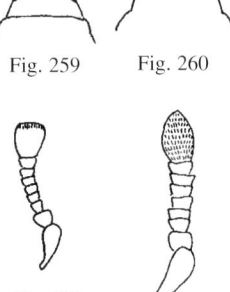

Fig. 261

Fig. 262

Genus *Pentarthrum* Wollaston

Four species of this genus, as at present constituted, are known. They have a wide distribution globally. Only one species has a permanent place on our faunal lists, but *P. elumbe* (Boheman) was included by Buck (1948; as *P. cylindricum* Wollaston) as a possible coloniser of the British Isles; this has not occurred. That species, one individual of which was imported into Manchester, has the pronotum with straight sides, almost parallel-sided, and broadest at extreme base.

- One species; a linear, pitchy to black cossonine, distinguished by the five-segmented antennal funiculus (cf. fig. 263) and simple elytra apicad. .... *huttoni*

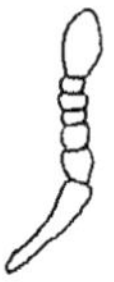

Fig. 263

Male rostrum shorter, broader (about twice as long as broad), more strongly and closely punctured, so duller, and more strongly curved; scrobe extending almost to rostral apex; antennae inserted in front of middle. Female rostrum longer, narrower (about 3 x as long as broad), less strongly and more remotely punctured, so more shining; scrobe terminating well before rostral apex; antennae inserted behind middle.

Generally synanthropic, occurring in houses, cellars and industrial premises, attacking a wide variety of hard- and softwoods. In floorboards, panelling, casks and plywood structures. The substrate is usually damp and infested with fungus, most often Coniophora puteana *(Hum, Glaser & Edwards, 1980). The early stages and aspects of the species' biology were described by Hammad (1955). Very local and not often met with (sometimes confused with the commoner* Euophryum confine*), but widely, if sporadically, distributed throughout the British Isles as far north as Lanark and E Lothian, though commoner in southern England. Local, but widely distributed in Ireland. Not widely distributed in Europe but recorded from France, Belgium, Germany and Denmark (introduced). Thought to be a native of Chile (R. T. Thompson pers. comm.)*

90

Genus *Euophryum* Broun

This genus, one of two representing the tribe Pentarthrini in the British Isles, contains three species, two native to New Zealand and the other to Chile (Thompson, 1989). The two New Zealand species are well-established here, with *E. confine* occurring very commonly. Both species are also recorded from central Europe (Folwaczny, 1983) but are not widely distributed there. As recent introductions to the British Isles neither species appears in Fowler (1891) or Joy (1932).

Key to species

Rostrum abruptly excised at base (best seen from below or obliquely from the side; fig. 264), rostrum somewhat dilated in oblique view; apex of antennal club rounded [generally lighter in colour, dark reddish; mainly in structural (converted) timber; 2.5-3.2 mm]. .... ***rufum***

Fig. 264

Male rostrum shorter, about twice as long as broad (fig. 264); pronotum narrower, 1.3 x as long as broad. Female rostrum longer, about 2.5 x as long as broad (cf. fig. 265); pronotum broader, about 1.24 x as long as broad.

In buildings, associated with old timber such as floor-boards, joists and panelling, especially where the timber is damp. Also occasionally in stored products (Buck, 1948; Hodge & Jones, 1995), though this habitat is not general. Introduced, scarce and sporadic, but widely distributed in England, Wales and Ireland; it has been found relatively frequently in London. No Scottish records. Native to New Zealand, where it exists as 'northern' and 'southern' forms; European examples belong to the northern form (Thompson, 1989). Introduced into Denmark and western Switzerland, but not otherwise known from the continent of Europe.

- Rostrum more shallowly excised at base (fig. 265), rostrum not dilated in oblique view; apex of antennal club bluntly acuminate [generally darker in colour; dark brown to pitchy; in dead wood outdoors as well as in structural timber; 2.5-3.6 mm]. .... ***confine***

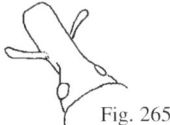

Fig. 265

Male rostrum duller, shorter and broader (2.0-2.1 x as long as broad; cf. fig. 264); pronotum more rounded at sides, broadest only a little behind middle. Female rostrum more shining, longer and narrower (2.7-2.8 x as long as broad; fig. 265); pronotum less rounded at sides and broadest well behind middle.

In all kinds of dead timber; well-established in the open and occurring in hedges, fallen trees and dead wood (often damp and rotten) generally. Associated particularly with wet and rotted wood, for example old floorboards, indoors; often in timber infested with the fungus Coniphora puteana. *It will feed on fungus-infested cardboard and paper (Hum, Glaser & Edwards, 1980). A frequent pest in buildings, especially in London (Hickin, 1968). The larva has been briefly described by May (1993). Introduced; now common throughout most of Great Britain, occurring as far north as Shetland. Probably widespread in Ireland, though recorded mainly from the north. Its spread through the British Isles since its first record in 1937 has been rapid and thorough. Also occurring as two forms in its native New Zealand, though these are not so distinct as those of its congener (Thompson, 1989). Introduced into Austria and Denmark.*

Genus *Pselactus* Broun

The sole British species in this genus is one of our most distinctive cossonines.

- One species; a dark brown, black or pitchy weevil with slightly but characteristically rounded elytra (fig. 266) and conspicuous pale upstanding setae. .... ***spadix***

Fig. 266

Male rostrum shorter, about 1.8 x as long as broad; antennae inserted nearer rostral apex, distance from insertion to apex about equal to rostral width at insertion. Female rostrum longer, about 2.4 x as long as broad; antennae inserted further from rostral apex, distance from insertion to apex about 1.5 x width of rostrum at insertion.

Predominantly, if not exclusively, a maritime species in the British Isles. In all kinds of timber by the sea shore, driftwood, piers, piles, wharves and groynes, attacking the timber of many hardwood and some softwood species; Oevering, Matthews & Pitman (2000) stated that it has a preference for the latter. Also recorded occasionally from structural timbers, such as floorboards, in maritime towns. Recently reported as causing significant structural damage to wharf timbers in southern England (Sawyer & Cragg, 1995). Often, perhaps usually, in timber which is immersed for longer or shorter periods in sea water. It has been estimated that adult weevils are able to withstand immersion for up to 7 hours a day; no doubt the body setae prevent, or delay, wetting, acting as an air-store. The weevils may also use air trapped in their tunnels in the inundated wood while submerged (Sawyer & Cragg, 1995). Not uncommon in suitable locations; recorded from most of the maritime vice-counties of England as far north as SE York, but not known from the NW coast (Oevering, Matthews & Pitman, 2000). Also recorded from Wales (Glamorgan) and Ireland (Co. Dublin), but not reported from Scotland or Isle of Man. Widespread in central Europe, where it is not apparently restricted to the coast. Several subspecies and forms have been described (Folwaczny, 1983), but some of these appear to be of doubtful value (Abbazzi & Osella, 1992). Introduced to, and established in, North and South America, Australia and New Zealand.

Genus *Caulophilus* Wollaston

Sixteen species of the genus are known (Kuschel, 1962), most having been recorded from North and Central America. The listing of the genus (and species) as part of the British fauna (Kloet & Hincks 1945; Pope 1977; Hodge & Jones 1995) is based on weak evidence.

- One species; a small (2.35-2.95 mm), large-eyed (fig. 267), pitchy-black weevil.. **oryzae**

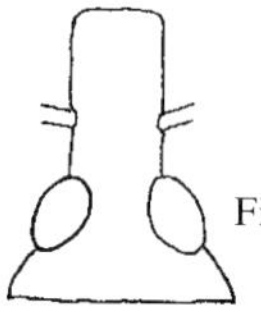

Fig. 267

Differences between the sexes apparently slight; the male rostrum is a little shorter than that of the female and the male tarsal claw-segment tapers distad.

A species of stored products which occurs indoors and under artificial conditions. Although it is included in the most recent checklists of our fauna, as noted above, Mound (1989) states: "The absence [present author's emphasis] of British records may be due to its close superficial resemblance to the species of Sitophilus". Any resemblance is indeed superficial and entomologists are unlikely to confuse the two genera. Occurrence and distribution in the British Isles therefore uncertain, though the species is known to be not infrequent in ship-borne cargo from abroad, particularly ginger. Formerly maintained in culture at the (then) Pest Infestation Laboratory, Slough. There are no British specimens in the National Collection (in The Natural History Museum). Not included as a Palaearctic species by Folwaczny (1973) but stated by O'Brien & Wibner (1982) to occur in Europe. Known from the Caribbean, Hawaii, North and Central America; a pest particularly in the southern USA (Mound, 1989).

Genus *Pseudophloeophagus* Wollaston

Our sole British species was formerly placed in *Caulotrupodes* and before that in *Caulotrupis*.

- *One species; a somewhat short and robust weevil, black to pitchy, shining, often with a brassy reflection, and with characteristically slender, reddish antennae (fig. 268).*
  .... ***aeneopiceus***

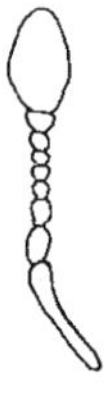

Fig. 268

Male rostrum shorter and broader, about 1.2 x as long as broad; antennae inserted at about 1/3 from apex. Female rostrum longer and narrower, about 1.5 x as long as broad; antennae inserted at about 2/5 from apex.

In woodland, open situations and occasionally synanthropic. In rotten wood, often by the sea, but also inland; recorded from old wine casks, posts and driftwood, and an uncommon and very minor pest of structural timber in buildings (Hum, Glaser & Edwards, 1980; as Caulotrupodes). *Also occasionally in woodland leaf litter. Not uncommon, but local and sporadic, often abundant where it occurs. Recorded widely in the British Isles: throughout England and Wales, southern Scotland, with an outlying record from the Hebrides; rare in Ireland. Very scarce and sporadic in Europe.*

Genus *Cossonus* Clairville

Two species of this genus inhabit the British Isles; a third occurs widely in central and southern Europe. The genus is very speciose in N. and Central America, but these species may not be congeneric with the European ones.

Key to species

Body more depressed, disc of pronotum and elytra flat; pronotum more coarsely punctured, punctures at base very large, deep, pit-like and often confluent, leaving an irregularly-sided, unpunctured, smooth, median basal line (fig. 269); pronotum regularly rounded at sides, broadest at middle; elytra with coarser, broader, circular strial punctures, striae nearly as broad as interstices; average size smaller, 4.4-5.1 mm. .... *linearis*

Fig. 269

Differences between the sexes very slight. Male rostrum slightly shorter, duller and more strongly punctured than female rostrum. ·

In open woodland, fens and parkland. In rotten wood, particularly of fallen trees, and under bark. It attacks especially Salix *and* Populus *and has also been found on* Pinus. *Mixed populations with C. paralellepipedus have been recorded (Nash, 1979; Drane, 1979). Local and not common. It has been recorded from a block of vice-counties in the south-east of England, from E Sussex to Buckingham and northwards to Cambridge, Huntingdon, E Norfolk and N Lincoln. Not recorded from Wales, Scotland, Isle of Man or Ireland, nor from northern and western England. Widespread in Europe, northwards to Denmark, southern Sweden and Lithuania.*

- Body less depressed, disc of pronotum and elytra slightly convex; pronotum more finely punctured, punctures at base little larger than those at apex and sides, shallow, neither pit-like nor confluent (fig. 270), median broad unpunctured line less obvious; pronotum less regularly rounded at sides, more abruptly contracted to base and broadest behind middle; elytra with finer strial, narrower linear punctures, striae much narrower than interstices; average size larger, 4.3-6.6 mm. .... *parallelepipedus*

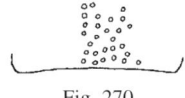

Fig. 270

Male rostrum from antennal insertion apicad shorter, distance shorter than, or equal to, width of rostrum at insertion; abdomen impressed at base, with yellow pubescence. Female rostrum from antennal insertion longer, distance greater than width of rostrum at insertion; abdomen simple.

In woods, parkland etc.; habitat similar to that of C. linearis. *In dead wood of mainly broad-leaved trees; species of* Quercus, Ulmus, Populus *and* Salix *are known hosts. As noted above, mixed colonies of our two* Cossonus *species have been recorded. Local, but widely distributed in southern England northwards to Cambridge, Huntingdon and Worcester with a more isolated record from Nottingham. Not recorded from Scotland or Ireland, but known from Wales (Glamorgan). Widely distributed throughout Europe, including all the countries of Fennoscandia.*

Genus *Rhopalomesites* Wollaston

Our sole species of this genus has often been placed in *Mesites*, which includes about twenty species, most of them occurring in the Atlantic Islands.

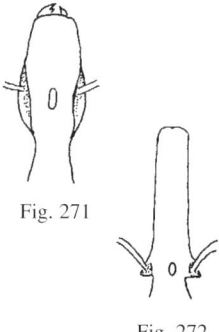

Fig. 271

Fig. 272

- One species; a large, distinctive cossonine (4.3-10.5 mm, exceptionally to 12.0 mm) with pronounced secondary sexual characters (figs 271, 272, 273); black to pitchy (habitus fig. 273). ....*tardyi*

Male rostrum dull, spade- or chisel-like distad, with convergent sides to apex, strongly narrowed behind antennal insertion, which is about one-third from apex (figs 271, 273); antennal scape longer, flagellum about 1.2 x its length. Female rostrum shining, much narrower than male, cylindrical, antenna borne on a sharp lateral tooth, rostrum excised behind antennal insertion, which is about one-fifth from base (fig. 272); antennal scape shorter, flagellum about 1.5 x its length. It should be noted that both Fowler (1891; pp. 389 n., 392-3) and Joy (1932) have transposed the characters of the rostrum in the two sexes, their males being actually females, and vice versa.

In woods, parkland and hedgerows, infesting the dead wood of a very wide range of mainly broadleaved trees. Traditionally associated especially with Ilex *(the 'Holly weevil'), but often found in species of* Fraxinus, Alnus, Crataegus, Quercus, Ulmus, Fagus *among others, including* Rhododendron *(Read, 1982). Aspects of the biology were described by Read (1982). Rather local, but not uncommon, and with a markedly western distribution in the British Isles, from E Cornwall to N Wilts., northwards in all the vice-counties of west Wales to north-west England and the west coast of Scotland as far north as Dumbarton and recorded from the Mid- and North Ebudes and Hebrides. There are more isolated records from E Sussex, E Norfolk and some of the Yorkshire vice-counties. Recorded from the Isle of Man and widespread and fairly common in Ireland. Possibly a native of the North Atlantic Islands (Read, 1982). Recorded from the Azores and elsewhere in Europe only from Norway.*

Genus *Rhyncolus* Germar

This genus is perhaps better known to British coleopterists as *Eremotes*. Five species, of about a dozen Palaearctic representatives, occur in central Europe but only one is found in the British Isles.

- One species; elongate, black to pitchy, legs generally somewhat lighter. .... *ater*

Differences between the sexes very slight. Male a little narrower than female.

In woods and parkland, usually associated with conifers. In dead wood, primarily of Pinus sylvestris, *but also recorded from timber of broad-leaved trees, especially in southern England, where the weevil is rare. Often found under bark of dead pines and in their stumps. Local, though widely distributed in Great Britain, but chiefly a species of the Caledonian pine forests, recorded from Stirling northwards to E Inverness. Also reported sporadically in southern England from S Somerset to E Kent and N Essex and again in W Suffolk, Nottingham and NE York Widely distributed throughout Europe, including the north.*

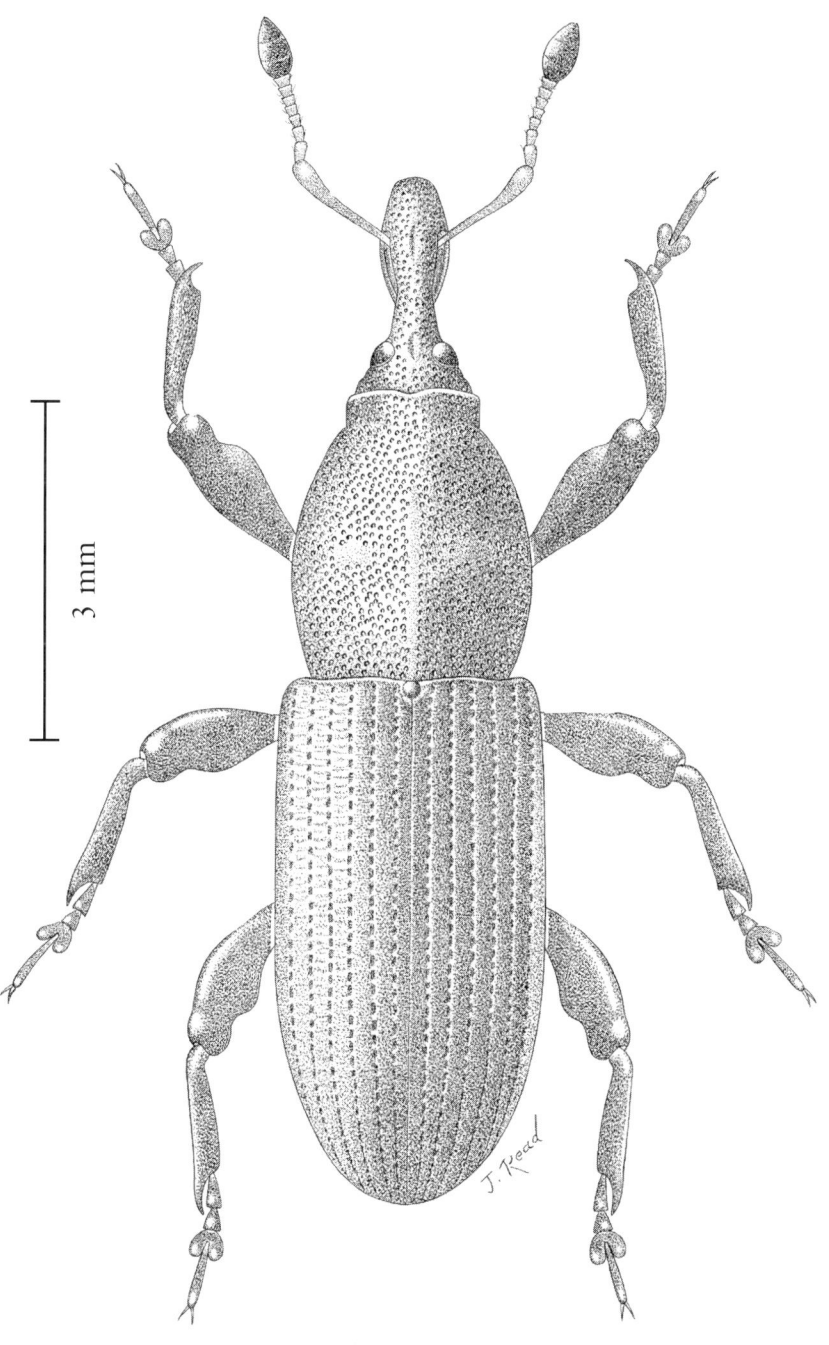

Fig. 273
Rhopalomesites tardyi

Genus *Phloeophagus* Schoenherr

As at present constituted this genus contains about 10 Palaearctic species, with five occurring in Europe. Of these only two have been recorded from the British Isles, one having been extinct for a century. They are typical cossonines, pitchy to black in colour, and fairly small.

Key to species

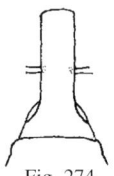

Fig. 274

More delicate, narrower and more elongate, elytra about 2.1 x as long as broad; slightly smaller on average, 2.6-3.4 mm; eyes almost flat (fig. 274); pronotum very indistinctly constricted subapically, 'collar' poorly developed (fig. 274) [species extinct in the British Isles].
....*gracilis* (Rosenhauer)

Differences between sexes slight. Male rostrum shorter (about twice as long as broad), parallel-sided. Female rostrum a little longer (about 2.6 x as long as broad), slightly constricted laterally between antennal insertion and apex.
In woods. Recorded from the dead or rotten wood of Fagus Crataegus, Ilex *and* Betula. *Not recorded in the British Isles during the twentieth century and assumed to be extinct. Older records exist from Surrey, Middlesex, W Gloucester, Warwick, Nottingham and S Lancaster. However, it is possible that some records are erroneous as the only specimens standing under the name* gracilis *in the National Collection (in The Natural History Museum) are* lignarius. *Records from Germany and Austria are also false (Folwaczny, 1983) and the species is known from southern, but not central or northern Europe. In France it occurs only in the south, so that its presence in the British Isles is anomalous. RDB Appendix.*

Fig. 275

- More robust, broader and less elongate (fig. 277), elytra 1.6-1.8 x as long as broad; slightly larger on average, 2.8-3.6 mm; eyes distinctly, though feebly, convex (figs 275, 277); pronotum distinctly narrowed subapically, with a clear, well-developed 'collar' (figs 275, 277)............ ***lignarius***

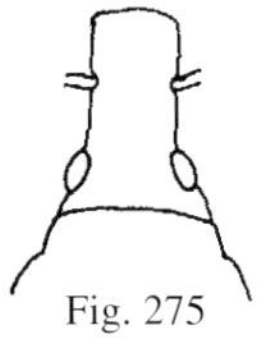

Sexual differences very slight. Male slightly narrower, pronotum a little less strongly rounded at sides. Female a little broader, pronotum more strongly rounded at sides.
In woods, forests and parkland. In dead wood of broadleaved trees, for example Quercus, Fraxinus, Ulmus, Fagus *and* Populus; *also in dead stems of* Hedera. *Often in the dead parts of hollow trees. Fairly common, and widely distributed throughout southern and midland England as far north as Nottingham and Chester. There is an old, isolated record from Scotland (Dumfries), and records from only three Welsh vice-counties, but the species is not known to occur in Ireland or the Isle of Man. Widely distributed in southern, central and northern Europe.*

Genus *Stereocorynes* Wollaston

Although this name has been used as the genus for our one species only recently, it was employed subgenerically by Fowler (1891) and so is reasonably familiar to British coleopterists.

- One species; a small, pitchy to black weevil (3.0-3.7 mm); elytra conspicuously subtruncate at apex; antennae characteristic (fig. 276). .... ***truncorum***

Fig. 276

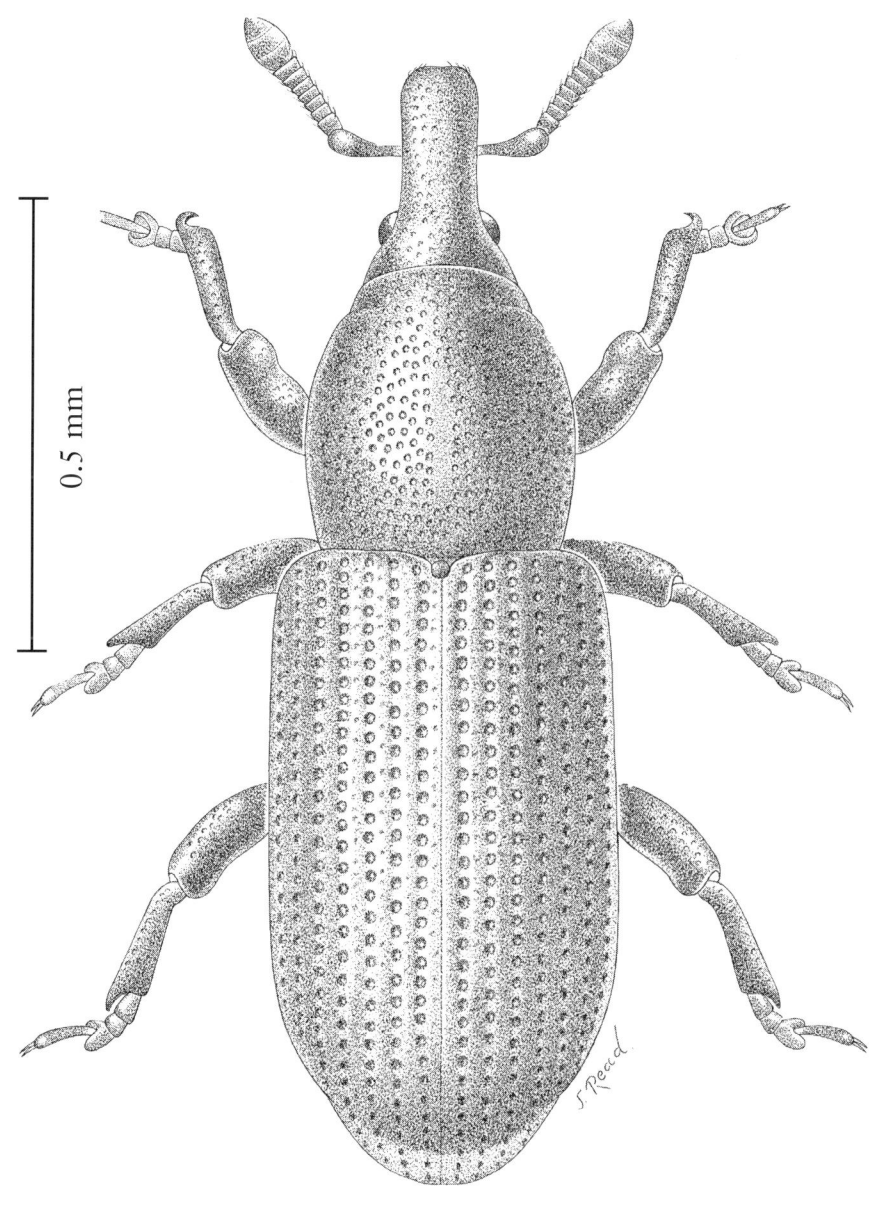

0.5 mm

Fig. 277
Phloeophagus lignarius

97

Male rostrum shorter and broader (about 1.3-1.4 x as long as broad at base), more strongly and abruptly constricted to apex. Female rostrum longer and narrower (about 1.8 x as long as broad at base), less strongly and more gradually tapered to apex.

In woods, pasture-woodland and ancient parks, occurring in the dead wood of a wide variety of broad-leaved trees, perhaps particularly Fagus. *Sometimes it occurs in company with* Phloeophagus lignarius. *Generally rare, but quite widely, if sporadically, distributed in southern England, from N Somerset to W Kent northwards to Leicester. No records from Wales, Scotland, Isle of Man or Ireland. Throughout central and southern Europe, northwards to Denmark, Latvia and southern Sweden.*

Genus *Macrorhyncolus* Wollaston

It is not certain that the recently recorded British species is correctly placed in this genus.

- One species; a small, delicate, red to pitchy weevil of which only three specimens have so far been found in the British Isles. .... *littoralis*

Male virtually indistinguishable from female on dorsal characters, but with a ventral depression and ventrites duller, with closer microsculpture, compared with female.

A littoral species associated with dead wood. Three British specimens, two taken in E Kent (Shakespeare Cliff, Aycliff, in a pitfall trap, June 1987), the other in W Kent (Dartford, on the tyre of a bicycle, 22 March 1988) (Welch, 1990). Possibly native to Australia, but well-established in New Zealand and recorded from South Africa, Chile and California.

Subfamily Cryptorhynchinae

This is a large and cosmopolitan subfamily, containing about 7000 species (Papp, 1979; Thompson, 1992). Many species are associated with dead wood, but others are stem-borers or feeders in fruits and seeds. Since at least the time of Wollaston (1860) stridulation has been recognised in representatives of the subfamily (Lyal & King, 1996). The subfamily is only moderately speciose in Europe, and in the British Isles there are a mere four resident species in two genera, classified in separate subtribes of Cryptorhychini (Alonso-Zarazaga & Lyal, 1999). Mango weevils, particularly *Sternochetus mangiferae* (Fabricius), are occasionally found imported into Britain with fruit.

Key to genera

Size large, 6.7-8.7 mm; scutellum large and obvious (fig. 278); apical third of elytra whitish, contrasting with darker basal two-thirds [on species of *Salix*, especially *S. viminalis* and *S. triandra* (Osiers)].
.... (subtribe **Cryptorhynchina**) *Cryptorhynchus*

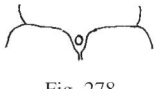

Fig. 278

- Size much smaller, 1.9-3.8 mm; scutellum small and inconspicuous, or invisible; apical third of elytra not whitish, elytra not strongly demarcated into two differently coloured regions [not usually on *Salix*].
(subtribe **Tylodina**) *Acalles*

Genus *Cryptorhynchus* Illiger

Only one species in this genus occurs in western Europe, although about six are known from the eastern Palaearctic. The genus is much more speciose in North America, where about 200 species occur. *C. harrisoni* Pool, described from Britain, has been identified as an introduced Australian species, *Pseudostoreus placitus* Lea, currently believed to be tychiine (Thompson, 1988).

-   One species; a large (6.7-8.7 mm), bicoloured, black-and-white weevil, the apical third of elytra unicolorous whitish, clearly demarcated from the basal two-thirds which are dark, predominantly black, but with a variable amount of intermixed white scales; interstices 3, 5 and 7, and pronotum, with tufts of raised scales. .... *lapathi*

Male rostrum shorter, more strongly punctured and duller, the apical portion from antennal insertion as long as antennal scape; first abdominal segment with a median longitudinal impression. Female rostrum longer, less strongly punctured and more shining, the apical portion from antennal insertion longer than antennal scape; first abdominal segment not impressed.

In fen carr, by the banks of rivers and other water-courses, and in damp woodland and osier beds. On species of Salix, *especially S.* triandra *and S.* viminalis, *larvae in the stems, forming swellings. A pest of these osiers, which are used for basket weaving (Smith & Stott, 1964), though not of cricket-bat willows* (S. alba var. caerulea) *on which it also occurs (Callan, 1939). Also recorded as attacking other species of* Salix, *and* Alnus *and* Betula *spp., either in the British Isles or in continental Europe. Adult weevils resemble bird droppings and are found predominantly in May and June. The life-cycle takes two years in the field, with adults not emerging from their breeding stems after eclosion in August until spring of the following year (Smith & Stott, 1964). In British Columbia the life cycle takes up to three years (Harris & Coppel, 1967). Widely distributed throughout the British Isles to S Aberdeen, but local and not common. Not in the Isle of Man, nor most of the Scottish Islands, though there is an outlying record from N Ebudes. Very local in Ireland. Very widely distributed throughout the northern and central regions of the Palaearctic and in North America.*

Genus *Acalles* Schoenherr

This is a large genus of generally small, inconspicuous, rather obscure species of unobtrusive habits. Adults frequently feign dead, tucking the legs, rostrum and antennae into the body, and may be passed over as small pieces of wood or debris. The genus is well-represented in such regions as southern Europe, the Atlantic Islands and North America. Thirty-five species have been recorded from France, but only three are known to occur in the British Isles. Much revisionary and descriptive work has been done recently on groups of the genus, particularly in central Europe, and the British representatives are perhaps in need of a critical review. The biology of species of *Acalles* is poorly known, but many species are associated with dead wood, particularly twigs, of deciduous trees.

Key to species

1   Upper surface with short, curved setae or crests of short, broad scales, but without conspicuous upstanding scales, legs and head with apressed scales but without upstanding ones; base of head clearly visible; pronotum transverse, less strongly contracted to base, base conspicuously broader than apex (figs 279, 280) [more generally ground-living]. **2**

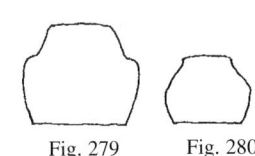

Fig. 279   Fig. 280

-   Upper surface and legs with conspicuous, broad, upstanding scales; base of head obscured and overlapped by upstanding scales of apex of pronotum; pronotum more strongly contracted to base, base little broader than apex (fig. 281) [more frequently arboreal]. ***misellus***

Fig. 281

Secondary sexual differences slight. Male rostrum a little shorter, apical part from antennal insertion less than twice length of antennal scape. Female rostrum slightly longer, apical part twice length of antennal scape, or nearly so.

In woods, at their edges, and in hedgerows; occasionally on isolated trees in pastures etc. Often found in association with dead Hedera *and* Crataegus. *Biology unknown; larvae probably in dead twigs and small branches. Generally common throughout southern and central England as far north as Derby, and in Wales and Isle of Man. Also in Scotland, from the Borders northwards to Dunbarton, though not recorded from the Highlands or Islands. Widespread in Ireland. However, it is probably under-recorded throughout the British Isles because of its obscure and cryptic habits. Patchily distributed in central and northern Europe; possibly confused with other species.*

2 Pronotum with a median longitudinal furrow, deeper and broader at base; pronotum somewhat abruptly constricted subapically, basal two-thirds less strongly convergent (fig. 282); elytra with conspicuous, but incomplete, raised interstices with crests of setae [small individuals often have elytral interstices less conspicuously raised than is the case in large specimens]; raised third interstice at elytral base especially prominent; average size larger, 2.5-3.6 mm; generally more strongly variegated in colour, especially prothorax. .... ***roboris***

Fig. 282

Secondary sexual differences not marked. Male rostrum slightly shorter, more strongly punctured, and duller. Female rostrum a trifle longer, more weakly and remotely punctured, and more strongly shining.

In deciduous woods, by beating Quercus *and other trees; often in leaf litter and on the ground in wooded areas. Larvae probably in dead wood, particularly twigs, of oak and perhaps other trees, but biology largely unknown. Local and not usually common but occasionally taken in numbers, particularly in woodland litter samples and pitfall traps. Sporadically recorded throughout England and Wales as far north as Durham, but with a very patchy vice-county distribution, and extending only into the Borders and Lothians in Scotland (Dumfries, Kirkcudbright, Roxburgh and Edinburgh). Rare in Ireland (Cos. Kerry, Carlow and Wicklow). Throughout southern and central Europe as far north as Denmark and southern Sweden and Norway.*

- Pronotum without a median longitudinal furrow; pronotum more evenly rounded at sides, basal two-thirds more evidently rounded and contracted to base (fig. 283); elytral interstices less strongly, but completely and regularly raised; without conspicuously raised third interstices at elytral base; average size smaller, 1.9-3.0 mm; generally less strongly variegated in colour, especially prothorax. .... ***ptinoides***

Fig. 283

Secondary sexual characters not marked. Male rostrum shorter, less than twice as long as antennal scape. Female rostrum longer, about twice as long as antennal scape.

In woods, generally on acid soils, and on heathland. Ground-living and not usually arboreal. Often found in moss and leaf litter and in pitfall traps. Biology unknown, but larvae probably in dead twigs of trees, especially Quercus, *and* Calluna. *Not uncommon, and widely distributed throughout England, Wales and southern Scotland, extending northwards to W Perth and Dunbarton. Also in Isle of Man; local and rather uncommon in Ireland. In western Europe only, extending northwards to Denmark and southern Sweden and Norway, but not recorded from eastern Europe.*

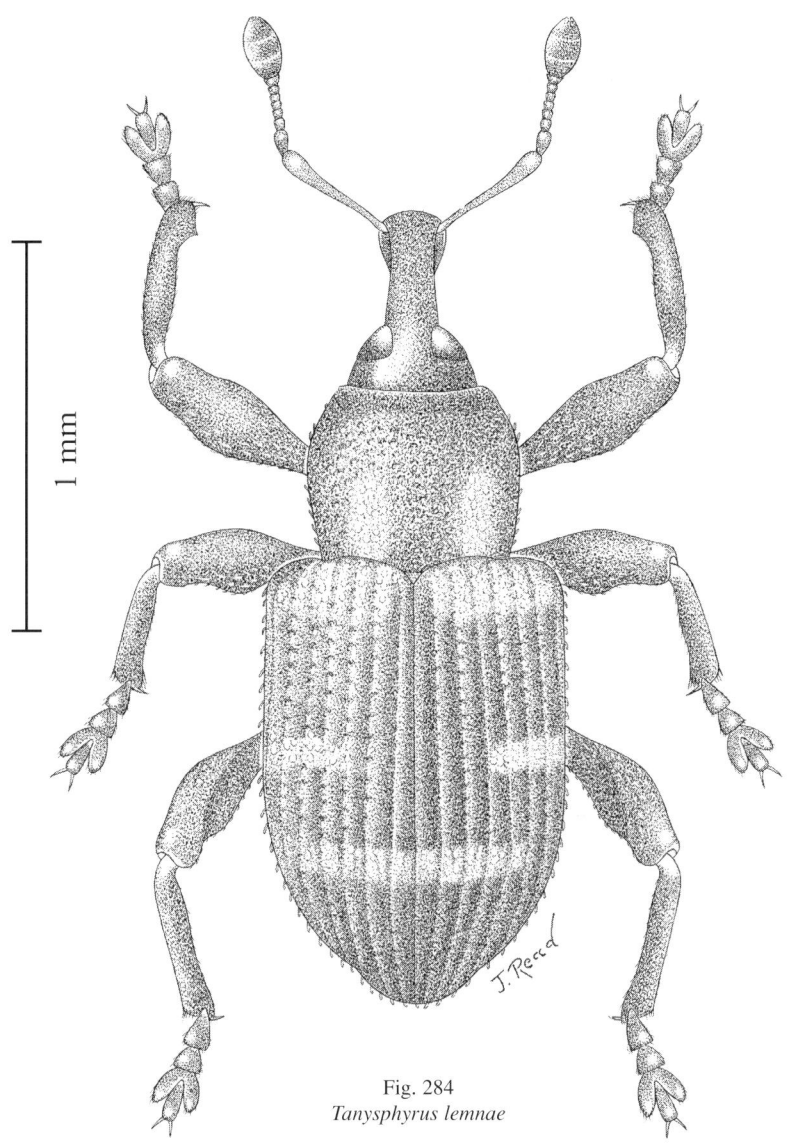

Fig. 284
Tanysphyrus lemnae

Subfamily Tanysphyrinae

The affinities of this group, now recognised as gonatocerous, are uncertain. Treated here as a subfamily, following Pope (1977), it contains three genera of which one is represented in Europe, including the British Isles.

Genus *Tanysphyrus* Germar

This genus includes four species, distributed in the Holarctic region. Two occur in Europe, but only one has been recorded from the British Isles. They are small weevils associated mainly with floating aquatic plants. *T. ater* Blatchley (= *makolskii* Smreczynski; O'Brien & Wibner, 1984) is associated with aquatic liverworts and is distinguished from our species by a stronger subapical tooth to the fore-tibia, darker legs and less robust rostrum.

- One species; a small (1.45-1.85 mm) weevil with reddish tibiae, characteristically short tarsal claws (figs 284, 285) and conspicuously large antennal clubs (fig. 284); elytra short, with an obscure pattern of pale scales (habitus fig. 284)... ***lemnae***

Fig. 285

Male rostrum slightly shorter and duller, antennae inserted nearer apex, at a distance subequal to rostral breadth at insertion. Female rostrum a little longer and more shining, antennae inserted further from rostral apex, distance to insertion greater than breadth of rostrum at insertion.

At the sides of ponds, canals, slow-flowing rivers and still and stagnant water-bodies generally. On species of Lemna, *range of species used as foodplants uncertain; recorded especially from* L. minor *in continental Europe. Adults crawl over the floating plants and also occur on mud at the sides and banks of water-bodies; they are unable to swim. Also recorded from* Calla palustris *in Europe. Larvae in the floating thalli (of* Lemna*). Common and often abundant in suitable locations. Distributed throughout England and Wales as far north as Mid-Lancaster and NE Yorks. Local, but probably under-recorded in Ireland. No records from the extreme north of England, Scotland or Isle of Man. Throughout Europe, including all the countries of Fennoscandia, to Siberia. Japan and North America.*

Subfamily Bagoinae

This subfamily is one of the several disparate groups formerly included in Erirhininae in the old, 'gonatocerous' sense (e.g. Pope, 1977). The world Bagoinae are currently being revised by Caldara and O'Brien. Among their findings are that *Hydronomus* (and the non-British *Dicranthus*), previously treated as good genera, should be synonymised with *Bagous* (Caldara & O'Brien, 1998). Accordingly, *Hydronomus* is accorded the status of a subgenus of *Bagous* in the following account, thus reverting to the treatment of Fowler (1891).

Genus *Bagous* Germar

This is a particularly interesting genus in a number of different respects. The species are mostly rare, with very patchy distributions, not only in the British Isles but throughout western and central Europe. They appear to be relict species, survivals from the cooler periods of interglacials and they have been adversely affected by human activities, particularly drainage and pollution. At least three species (*petro, binodulus* and *robustus*) have almost certainly become extinct in Britain during historic times, and doubts must also exist about *diglyptus*. All species are semiaquatic (*B. lutulosus* and probably *B. diglyptus* the least so) and are well-adapted for life in fresh water, particularly in having well-developed plastron respiration (Hinton, 1976). Little has been recorded on the host plants of the species in the British Isles, but this deficiency is partially remedied by continental work.

Caldara & O'Brien (1998) point to anomalies in the traditional subgenera of *Bagous*, using species-groups to indicate affinities between species, but subgenera are retained, informally, here, for reasons of continuity.

The status of *Bagous arduus* Sharp has been investigated in detail by Hammond (1998). Although his findings are not completely conclusive, it is most probable that Sharp's holotype (the only unambiguous exemplar of his species) is an aberrant or damaged specimen of *B. longitarsis* Thomson. This identification was accepted by Caldara & O'Brien (1998) and is adopted here.

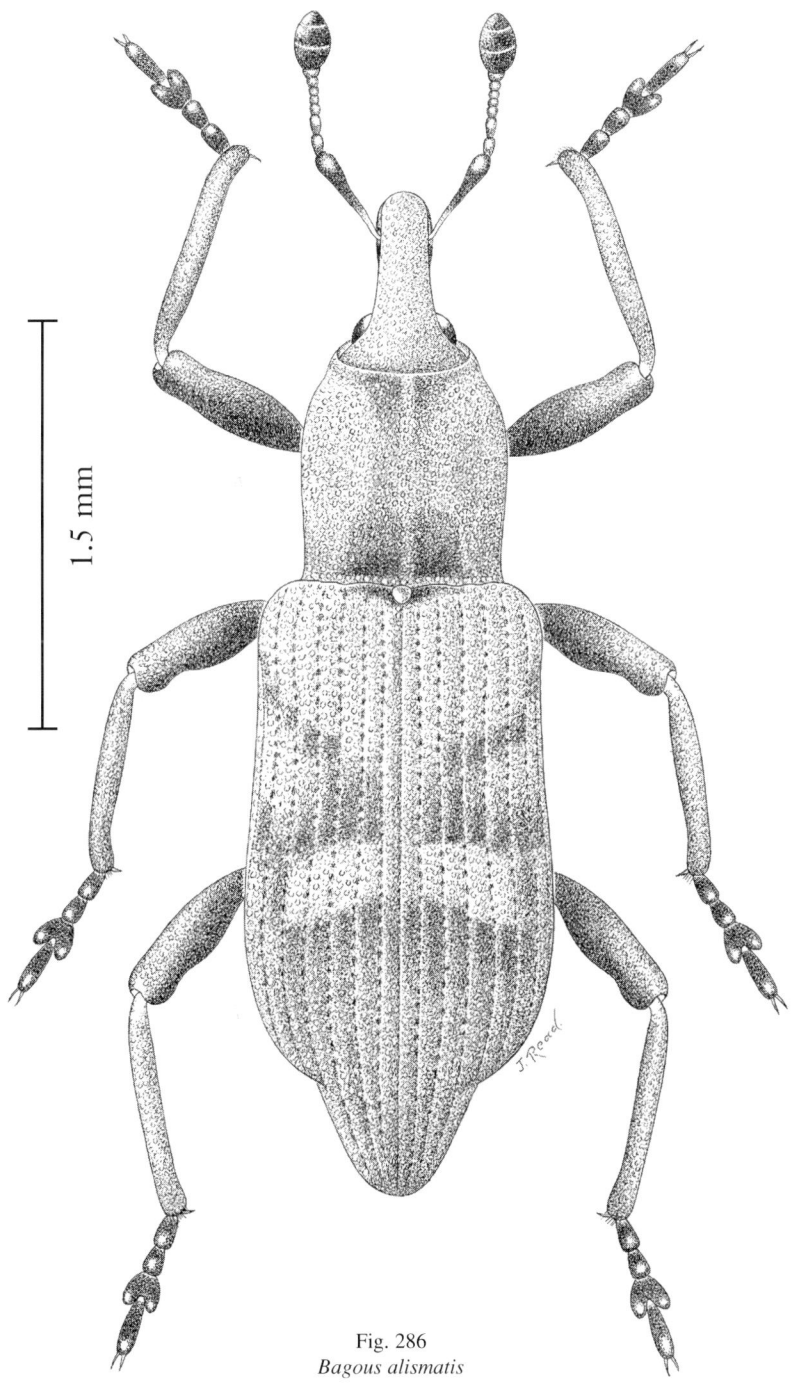

Fig. 286
Bagous alismatis

Bagous claudicans Boheman, listed as synonymous with *B. collignensis* in British literature (e.g. Pope, 1977), has been shown to be a good species (Dieckmann, 1990; Caldara & O'Brien, 1998). It is considerably smaller and has a different host plant (*Equisetum*) compared with *B. collignensis*. It is not known to occur in the British Isles.

Species of *Bagous* are difficult to identify, and some are distinguishable on small and comparative characters; the shape, proportion and size of the tarsal segments are particularly important features. Many of the records in the older British literature are inaccurate and there has been considerable confusion in nomenclature. The work of Dieckmann (1964), followed recently by the more extensive study of Caldara & O'Brien (1998), has established a reliable basis for the identification of the European species of the genus and this account and key are largely based on their findings.

Key to subgenera and species

1   Tarsi glabrous, or almost so, with at most a few sparse, long, fine setae; anterior margin of prosternum deeply excavate, the sides strongly keeled (fig. 287); generally on semi-aquatic plants other than *Alisma*.................................... **2**

Fig. 287

-   Tarsi thickly clothed with fine, short, pale setae or silvery pubescence; anterior margin of prosternum only weakly concave, and without lateral keels (fig. 288); on *Alisma plantago-aquatica* [elongate, with long legs, 2.7-3.5 mm]. One species (habitus fig. 286). ....(***Hydronomus***) ***alismatis***

Fig. 288

Differences between the sexes slight. Male rostrum slightly shorter than female rostrum.
In ditches, dykes, canals, ponds and other standing water-bodies. On Alisma plantago-aquatica, *and recorded in continental Europe also on* Sagittaria sagittifolia. *Larvae in blotch-mines on the leaves. Rather local, though often in numbers when found (the commonest British bagoine), and widely distributed throughout England, Wales and Ireland, extending northwards to Lanark and Edinburgh. Not recorded from Isle of Man. Throughout Europe and extending to Asia Minor and eastern Siberia.*

2   Antennal club oval, strongly developed, first segment dull, setose, subequal in length to remainder of club (fig. 289). **3**

-   Antennal club slender, weakly developed, first (basal) segment glabrous, shining, twice as long as remainder of club (fig. 290). [body very short, elytra rectangular, parallel-sided; pronotum strongly rounded at sides, broadest in front of middle, basal two-thirds with conspicuous pale scales which contrast with black elytra; size small, 2.7-2.9 mm; almost certainly extinct in the British Isles]. .... (***Ephimeropus***) ***petro***

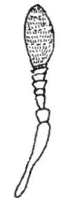

Fig. 289

Fig. 290

Rostrum slightly shorter in male than female.
In acid bogs; associated with Utricularia *spp., especially* U. vulgaris. *Details of biology unknown. Recorded only from Askham Bog, Mid-W Yorks, and almost certainly extinct. The drying out of the locality, presciently predicted by Fowler (1881), was probably the cause of its demise. According to Fowler & Donisthorpe (1913) only two genuine specimens were ever taken (one by Fowler and one by Hey) and, if this is so, both are in the National Collection (in The Natural History Museum). However Hey (1895) records taking two specimens and, as these were determined by David Sharp and the species is extremely distinct, it may be that at least three British specimens were collected. Very rare, but quite widely distributed, in northern and central Europe. RDB Appendix.*

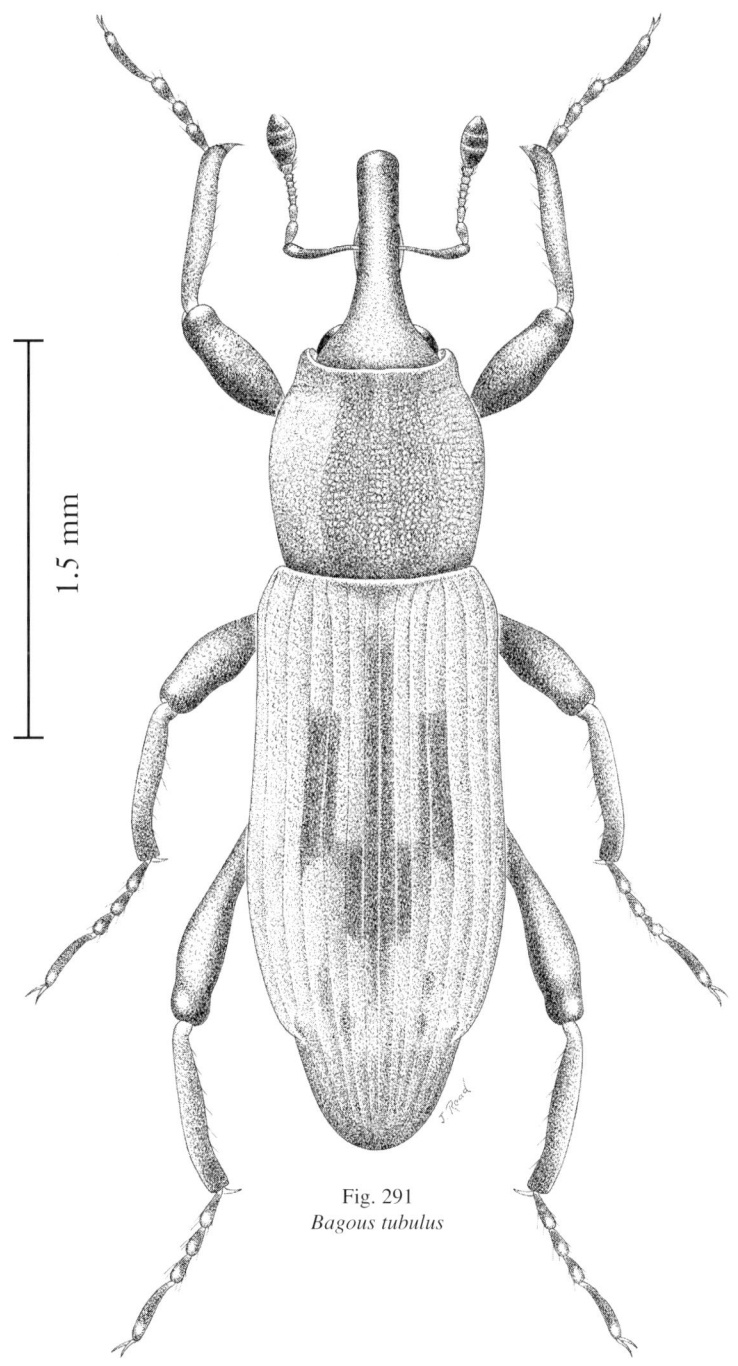

Fig. 291
Bagous tubulus

1.5 mm

3   Body long, cylindrical; pronotum almost as broad at its widest point as elytra at base (fig. 292); elytra twice as long as broad, humeri weakly developed; antennae inserted behind middle of rostrum [segments 1-3 of hind tarsus much longer than broad; length 2.6-3.6 mm] (habitus fig. 291)... *(Cyprus) tubulus*

Fig. 292

Male rostrum shorter; antennae inserted at about 2/3 from apex. Female rostrum longer; antennae inserted at about 3/4 from apex.

On the banks of dykes, reservoirs and other water-bodies, often with a clay substrate. Graminivorous, on a range of aquatic and semi-aquatic grass species. Recorded from Glyceria fluitans, G. plicata *(= notata) and* Aleopecurus aequalis *(= fulvus) in continental Europe; all three are common British grasses. Adults also reported as feeding on a species of* Callitriche *in France. Biology unknown. Rare, or very local, but sometimes common where found. Recorded from seven vice-counties in south-east England. Widely distributed in western and central Europe. RDB2.*

-    Body shorter and more squat; pronotum much narrower than elytra (fig. 293), elytra shorter, at most 1.9 x as long as broad; humeri more strongly developed; antennae inserted at or in front of rostral mid-point............................... **4**

Fig. 293

4   Segment 3 of all tarsi not broadened, only as broad as segment 2 (fig. 294) [if segment 3 somewhat broadened it is always longer than broad]......................... (**Bagous** s.str.) **5**

-    Segment 3 of all tarsi broader than segment 2, mostly as long as broad, cordate or oval (fig. 295)............. (**Abagous**) **18**

Fig. 295

Fig. 294

5   Dorsum dull, generally brown, covered with fine granules or finely tuberculate, tubercules often papillate, especially clear on pronotum. **6**

-    Dorsum shining, as if varnished, whitish-grey, smooth, scales of elytra and pronotum flat, not papillate [segment 7 of funiculus approximated to club, strongly transverse and bearing long setae (fig. 296); length 2.7-4.7 mm; in brackish places].. **argillaceus**

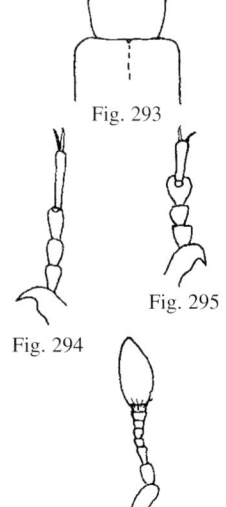

Fig. 296

Secondary sexual differences slight. Male rostrum slightly shorter, apical portion duller; antennae inserted slightly nearer apex. Female rostrum slightly longer, apical portion a little more shining; antennae inserted slightly further from apex.

A not infrequent form with brown-mottled elytra is f. inceratus *Gyllenhal.*

In brackish ponds and ditches on or near the coast (found in saline places inland in continental Europe). Biology and hosts unknown, though grasses have been suggested. Very local and generally rare, though occasionally in numbers when found. Recorded from S Hants (formerly), E and W Kent, S Essex and W Norfolk, the last record perhaps requiring confirmation. Though quite widely distributed in central Europe as well as on Atlantic coasts it has declined markedly in the last eighty years, particularly in inland areas (Lucht, 1987). RDB2.

6   Third segment of all tarsi longer than broad (fig. 294)........... **7**

-    Third segment of fore- and mid- tarsi as broad as or broader than long (fig. 297). **15**

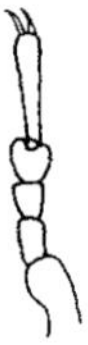

Fig. 297

7 Posterior third of elytra with a conspicuous node, swelling or protuberence on fifth interstice (figs 298, 299); larger species, 4.0-6.0 mm. **8**

- Posterior third of elytra without, or with a weakly developed node, swelling or protuberence on fifth interstice; smaller species, 2.3-3.6 mm. **9**

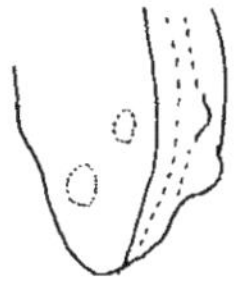

8 Posterior third of elytra with nodes, swellings or protuberences on both third and fifth interstices, subequal in size (node on third interstice at about 1/3, that on fifth about 1/5, from apex, fig. 298) [4.0-5.5 mm; on *Stratiotes aloides*, almost certainly extinct in the British Isles].
.... ***binodulus***

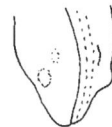

Fig. 298

Male rostrum a little shorter, more robust and slightly more shining; antennae inserted nearer apex of rostrum, apical portion about as long as rostral width at insertion. Female rostrum slightly longer, less robust and duller; antennae inserted nearer rostral base, apical portion evidently longer than rostral width.

In fens, ditches and broads, but this gives a false idea of the species' occurrence in Britain as only four specimens have been reliably recorded here (Newbery, 1902). On Stratiotes aloides, the larvae feeding on, or in, the leaves (Dieckmann, 1964). The only British locality recorded in any detail is Horning Fen, E Norfolk, where a specimen was taken in 1861. In central and the southern part of northern Europe, but it has declined in abundance in many areas. Listed as RDB+ (believed extinct).

- Posterior third of elytra with a node, swelling or protuberence on fifth interstice only (fig. 299) (if third interstice with a slight swelling this always smaller and weaker than protuberence of third interstice) [4.2-5.9 mm; on *Butomus umbellatus*]... ***nodulosus***

Fig. 299

Secondary sexual differences slight. Rostrum very slightly shorter and more robust in male than female.

In ditches, dykes, ponds and water-bodies with little or no flow. On Butomus umbellatus, larvae (and pupae) in flowering and vegetative stems. Rare and very local, with few recent records. From S Somerset eastwards to E Kent and northwards to Glamorgan (doubtful), Huntingdon (formerly) and E Suffolk, but recorded from fewer than 12 vice-counties. The southern part of northern, and central Europe to northern Italy, but recorded less frequently than formerly. RDB1.

9 Elytral striae fine, often continuous, with small, often obscure, narrow punctures (if striae somewhat broader then with narrower tarsi and pronotum only weakly rounded at sides). **10**

- Elytral striae coarse and broad, punctures discrete, large (fig. 300); [pronotum almost cordate, broadest in front of middle, strongly contracted in basal quarter, without a median longitudinal furrow; 2.5-3.4 mm]. .... ***limosus***

Fig. 300

Secondary sexual differences not marked. Male rostrum slightly shorter, antennae inserted nearer apex, apical portion a little shorter than scape. Female rostrum slightly longer, antennae inserted further from apex, apical portion about as long as scape.

In ponds, slow-flowing ditches and dykes, fens and bogs. Associated with Potamogeton spp.; larval feeding sites not known. Some aspects of the biology were described by Copestake (1999). The second-most widely distributed and least rare species of the genus in the British Isles. In most of the vice-counties of southern England, from W Cornwall eastwards to E Kent and northwards to Buckingham, E Suffolk and E and W Norfolk. Not recorded from the midland vice-counties, Wales or Scotland but present in Cheshire, some of the Yorkshire vice-counties and Cumberland. Scarce and local in Ireland (N Co. Kerry and Co. Clare). Widespread in central and northern Europe eastwards to Iran.

107

10 Body narrow; elytra more elongate, 1.6-1.9 x as long as broad (fig. 301); pronotum only slightly narrower (0.75-0.82 x) than elytra. **11**

- Body broader; elytra less elongate, 1.4-1.5 x as long as broad (fig. 302); pronotum evidently narrower (0.60-0.65 x) than elytra. **12**

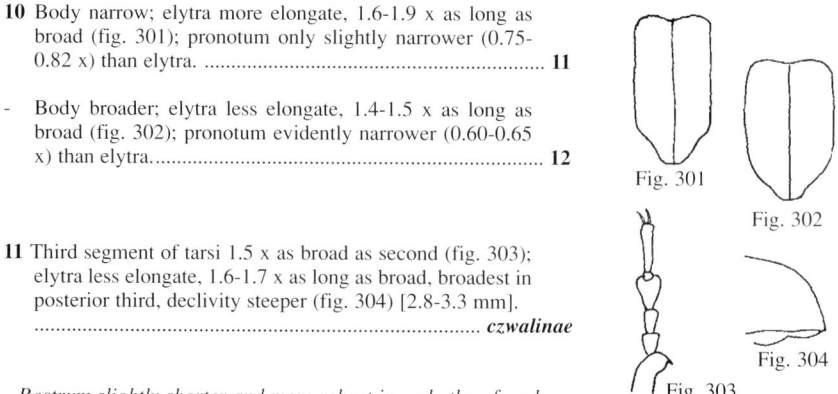

Fig. 301

Fig. 302

Fig. 303

Fig. 304

11 Third segment of tarsi 1.5 x as broad as second (fig. 303); elytra less elongate, 1.6-1.7 x as long as broad, broadest in posterior third, declivity steeper (fig. 304) [2.8-3.3 mm]. .... ***czwalinae***

Rostrum slightly shorter and more robust in male than female.

In small ponds and bogs, often with a clay substrate. Host plants and biology unknown (aquatic grasses have been suggested). Extremely localised and rare, known only from a few sites in the New Forest, S Hants. Quite widely distributed in northern and central Europe eastwards to northern Italy, but very rare throughout its range. RDB1.

- Third segment of tarsi only as broad as, or scarcely broader than, second (fig. 305); elytra narrower, 1.7-1.9 x as long as broad, broadest just behind base or with sides parallel, declivity shallower (fig. 306) [2.3-3.6 mm]. .... ***tempestivus***

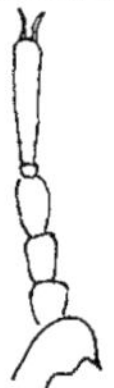

Fig. 306

Fig. 305

Male rostrum a little shorter, antennae inserted nearer apex, apical portion of rostrum about as long as antennal scape. Female rostrum a little longer, antennae inserted further from apex, apical portion of rostrum longer than antennal scape.

A variable species. In ponds, ditches, dykes; in standing or slow-flowing water-bodies generally. Also in wet grasslands. Possibly polyphagous; larvae have been recorded on Ranunculus *(e.g. R. repens) and adults have been found in association with* Potamogeton *and* Carex *spp., and with* Sagittaria sagittifolia. *Local, one of the least rare of the British species of the genus. Widely distributed throughout England and Wales to Mid-W York. Not recorded from Scotland, Isle of Man or Ireland. Widely distributed in Europe to Siberia.*

Fig. 307

12 Hind tarsi slender, longer, much more than half as long as tibia, each tarsomere more than twice as long as broad; suture separating basal portion of claw joint clearly visible (fig. 307); pronotum less strongly constricted subapically, pronotal 'collar' weakly developed; white spot on third elytral interstice usually conspicuous [tibial mucrones long and conspicuous (fig. 308); length 3.2-3.6 mm]. ....***subcarinatus***

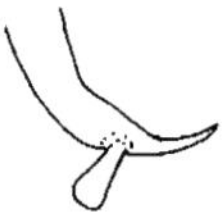

Fig. 308

Male rostrum slightly shorter and more robust, apical portion (from antennal insertion) about as long as broad at widest point. Female rostrum a little longer and less robust, apical portion longer (about 1.2 x) than maximum breadth. But secondary sexual differences not marked.

In dykes and ditches and other slow-flowing or standing water-bodies. Associated with Ceratophyllum submersum *but probably occurring also on other aquatic plants; larval feeding sites and other biological details unknown. Very local and usually rare. Recorded from seven vice-counties in southern England from N Somerset to E Kent northwards to Cambridge. Not in Wales, Scotland, Isle of Man or Ireland. Europe, though not the far north, Caucasus eastwards to Turkestan, North Africa.*

- Hind tarsi more robust, shorter, at least tarsomeres 2 and 3 less than twice as long as broad; basal portion of claw joint smaller, retracted into third tarsomere, visible only with difficulty (fig. 309); pronotum strongly constricted subapically, pronotal 'collar' strongly developed; white spot on third elytral interstice, if present, inconspicuous or merging with general pattern.. **13**

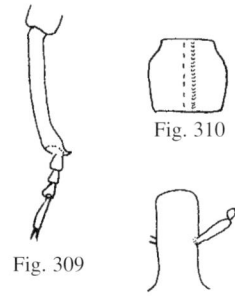

Fig. 310

13 Pronotum with a wide, deep median longitudinal furrow, as broad as an elytral interstice (fig. 310); pronotum broadest in anterior third, sub-cordate (fig. 310); antennal scape only as long as rostral width (fig. 311) [2.6-3.2 mm]. .... ***brevis***

Fig. 309

Fig. 311

Secondary sexual differences slight. Male rostrum a little shorter, and duller in apical half compared with female rostrum.

In and around ponds, bogs and fens. The host plant is Ranunculus flammula *in continental Europe (Cuppen & Heijerman, 1995) and probably also in Britain. Larvae on leaves and stems. Very rare and localised, known only from the New Forest, S Hants, and Surrey in England and not recorded from Wales, Scotland or the Isle of Man. Known from one site in the Burren, Co. Clare, Ireland, where it has been taken in some numbers. In central Europe eastwards to Siberia and in the southern part of northern Europe; very rare throughout its range. RDB1.*

- Pronotum without a wide median longitudinal furrow, at most with a fine, narrow line or groove which is not as wide as an elytral stria and much narrower than an interstice; antennal scape longer than rostral width (fig. 312)... **14**

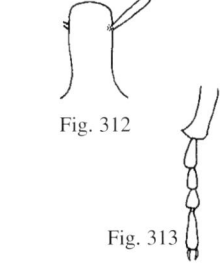

Fig. 312

14 Hind tarsi finer and slimmer, the second tarsomere longer than broad (fig. 313); pronotum more finely granulate (never with a median longitudinal groove); elytral disc flatter [2.4-2.7 mm]. .... ***longitarsus***

Fig. 313

Secondary sexual differences slight. Male rostrum a little shorter, more robust and duller compared with female rostrum.

In ponds, ditches, dykes and other slow-moving or standing water-bodies. Associated with Myriophyllum *spp., but larval hosts and biology unknown. This is one of several species in the genus which has been misidentified in the past and its relationship to* B. collignensis *is uncertain, the two taxa varying considerably throughout their ranges (Dieckmann, 1964). The distribution of* B. longitarsus *in Britain is unclear. Reliably recorded from only the New Forest, S Hants, and Romney Marshes, E Kent; other records require confirmation. A rare and very localised species. RDB1.*

- Hind tarsi coarser and shorter, the second tarsomere not longer than broad (fig. 314); pronotum with coarser granules (usually without a median longitudinal groove); elytral disc slightly curved and raised. .... ***collignensis***

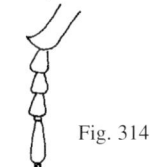

Fig. 314

Secondary sexual differences slight. Male rostrum a little shorter and more strongly curved, apical half slightly duller. Female rostrum somewhat longer and straighter, apical half a little more shining.

In ponds and ditches and on the banks of slow-flowing rivers. In continental Europe the hosts are Myriophyllum *spp. Local and generally rare, but widely distributed. In southern England from S Devon eastwards to W Kent and northwards to Cambridge. More isolated records from Cumberland and Kintyre; and from Cos. Westmeath, Armagh and Antrim, Ireland. Not reported from Wales or Isle of Man. Widespread and not exceptionally rare over most of Europe except the extreme north and extending to Anatolia. RDB3.*

15 Pronotum with an evident median longitudinal furrow, usually as wide as an elytral interstice, not always entire (cf. fig. 315, but narrower and not always continuous). **16**

- Pronotum without a median longitudinal furrow, occasionally with a fine, faint median line, always narrower than an elytral interstice. **17**

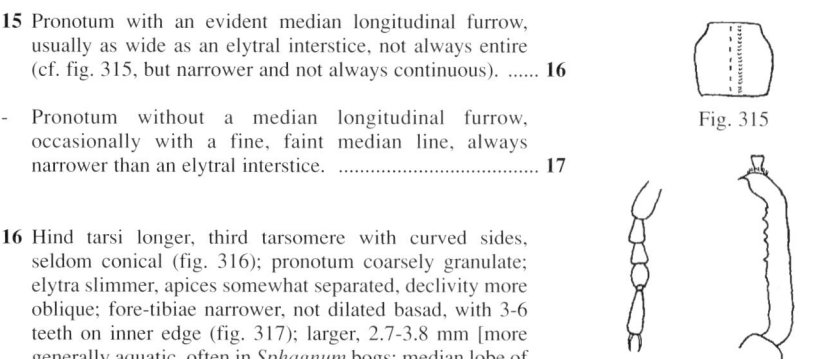

Fig. 315

16 Hind tarsi longer, third tarsomere with curved sides, seldom conical (fig. 316); pronotum coarsely granulate; elytra slimmer, apices somewhat separated, declivity more oblique; fore-tibiae narrower, not dilated basad, with 3-6 teeth on inner edge (fig. 317); larger, 2.7-3.8 mm [more generally aquatic, often in *Sphagnum* bogs; median lobe of male markedly asymmetrical]. .... *frit*

Fig. 316

Fig. 317

Male rostrum slightly shorter, antennae inserted nearer apex, apical portion of rostrum little longer than its width at insertion; apices of elytra more widely separated. Female rostrum slightly longer, antennae inserted further from apex, apical portion of rostrum about 1.5x as long as width of rostrum at insertion; apices of elytra closer together.

In Sphagnum *bogs. The host abroad, and almost certainly also in Britain, is* Menyanthes trifoliata. *Pupae have been sieved from* Sphagnum *in company with adults. Formerly regarded as one of our rarest species in the genus, being recorded only from the New Forest, S Hants and, not recently, Dorset and E Norfolk, though also known elsewhere from subfossil records. Recently discovered, partly by systematic sampling, to be widespread in Wales (Carmarthen, Cardigan, Caernarvon and Anglesey). Northern and central Europe, extremely rare; comparatively much rarer than in Great Britain. Listed as RDB1 (Shirt, 1987), but on recent information proposed for reclassification as RDB3 (Hyman & Parsons, 1992).*

- Hind tarsi shorter, third tarsomere with straight sides, conical or rectangular (fig. 318); pronotum more finely granulate; elytra more compact, apices not separated, declivity steeper; fore-tibiae broader, dilated basad, smooth on the inner edge (though with some outstanding setae) (fig.319); smaller species, 2.2-2.8 mm [in dryer situations than most *Bagous* spp., on *Juncus*]. .... *lutulosus*

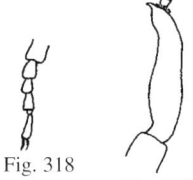

Fig. 318

Fig. 319

Secondary sexual characters not marked. Male rostrum slightly shorter, with antennae inserted a little closer to the apex, compared with female rostrum.

In damp situations in arenaceous areas, particularly those subject to periodic flooding on poorly-draining soils. Also occasionally in water. Associated with species of Juncus, *particularly* J. bufonius, *but biology unknown. Local and generally rare, though sometimes in numbers where found. Mainly in the southern coastal counties of England from Isle of Wight to E Kent and sparingly northwards to SE Yorks. Recorded from Wales (Glamorgan) but not from Scotland, Isle of Man or Ireland. Throughout Europe.*

17 Elytra shorter, about 1.3 x as long as broad (Fig. 320); tibia broader; tarsal segments shorter; colour of dorsum usually uniform grey [2.0-3.2 mm]. .... *diglyptus*

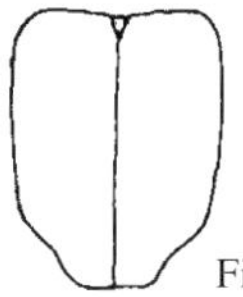

Fig. 320

Little information on secondary sexual differences.

One of the rarest, least well-known and most obscure of the British species of the genus. In continental European (Sweden and Germany) it is associated with both dry and moist situations, and the host is Saxifraga granulata. *The two original British examples were taken at Burton on Trent, Derby, in flood refuse from the R. Trent and from a wall nearby (Champion, 1879). Apart from these specimens (one of which is in the National Collection (in The Natural History Museum) and is certainly* diglyptus *as currently understood) there are records from E and W Suffolk and Sutton Broad, E Norfolk. However, specimens recorded from the last locality are* B. frit *(Blair, 1935b; confirmed for*

110

a specimen standing under the name diglyptus *in the National Collection). Two specimens collected from the River Gipping, E Suffolk in 1895 and 1897 are respectively in the Ipswich and Cambridge University Zoology Museums and are* diglyptus *(a very distinctive species). Central and southern Europe, extending to the southern part of the north of the continent. RDB1.*

- Elytra longer, about 1.5 x as long as broad; tibia narrower; tarsal segments a little longer; dorsum dark, with grey spots [see couplet 14].. ***collignensis***

18 Elytra broader, 1.3-1.45 x as long as broad, sides straighter, declivity steeper [there are good colour photographs of three of the species treated here (excluding *glabrirostris*) in Palm (1999)].. **19**

- Elytra narrower, 1.5-1.65 x as long as broad, sides rounded, declivity not so steep, but extended into a beak-shaped process, i.e. elytra strongly impressed at sides subapically (fig. 321); [white spot behind centre of third elytral interstice wanting or joined to other pale markings in a merged pattern; pronotum with a shallow median longitudinal furrow and an evident transverse furrow in anterior quarter, sides parallel to longitudinal furrow then convergent anteriad; tibiae and tarsi reddish brown, femora dark brown; large species, 3.5-4.5 mm]. .... ***lutosus***

Fig. 321

Secondary sexual differences slight. Male rostrum a little shorter, more strongly punctured, and duller, compared with female rostrum.

In dykes, ditches, ponds, lakes and other slow-flowing or standing water-bodies. Associated with Sparganium erectum *(= ramosum) in continental Europe; biology not known. Very local and rare, with recent records only from Ainsdale, S Lancaster. There are older records from N and S Hants, Surrey, W Norfolk and Leicester. Not recorded from Wales, Scotland, Isle of Man or Ireland. Rare in Europe; to Caucasus and Turkestan. RDB1 (proposed).*

19 Elytra without a V-shaped depression, occasionally with a shallow, obscure sub-basal depression not in shape of a V; pronotum narrowed anteriad and slightly at base, feebly rounded at sides; tarsi shorter, third tarsomere broader, cordate (fig. 322). **20**

- Elytra with a V-shaped depression on disc in anterior third (fig. 323); pronotum narrowed anteriad, subparallel at base, sides almost straight; tarsi longer, third tarsomere elongate and narrower (fig. 324) [3.5-4.0 mm]. .... ***puncticollis***

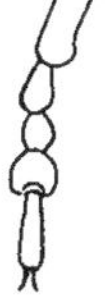

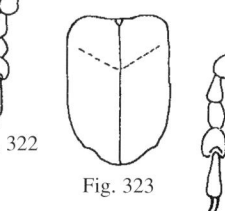

Fig. 322

Fig. 323

Fig. 324

Secondary sexual differences very slight. Male rostrum slightly shorter, and more strongly punctured, compared with female rostrum.

In ponds, dykes, ditches and other small water-bodies with no or slow flow. Evidence on host plants scarce, but thought to be polyphagous, associated with various aquatic plants in continental Europe, including Stratiotes aloides, Elodea canadensis *and* Hydrocharis morsus-ranae. *Biology not known. Very local and rare; recorded from Isle of Wight (perhaps requiring confirmation), N Hants, E Sussex, W Kent and Surrey. Not reported from midland or northern England, Wales, Scotland, Isle of Man or Ireland. Rare in Europe, but widely distributed except in the far north. RDB1.*

20 Pronotum more evidently rounded at sides, larger in proportion to elytra (fig. 325); elytral declivity less steep; tarsi reddish brown to black, always darker than apical quarter of tibia. **21**

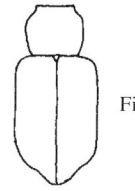

Fig. 325

111

- Pronotum only slightly rounded at sides, occasionally almost parallel-sided, with fine median longitudinal channel, smaller in proportion to elytra (fig. 326); elytral declivity steeper; femora generally dark; tibiae reddish brown, apical quarter and tarsi uniform rust-red; [antennae rust-red, scape sometimes darker apically, club blackish; 2.5-3.3 mm]. .... *glabrirostris*

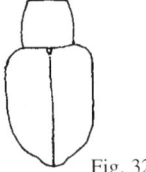

Fig. 326

Differences between the sexes slight. Male rostrum more strongly punctured, apical portion (from antennal insertion) slightly shorter. Female rostrum less strongly punctured, apical portion a little longer.

In a variety of freshwater biotopes, including ponds, lakes, ditches, dykes, fens, bogs, etc. Believed to be polyphagous, records of plant hosts in continental Europe include Ceratophyllum submersum, *and* Stratiotes aloides *on which larvae have been found feeding under water. Very local and generally uncommon, but widely distributed. From Isle of Wight and S Hants eastwards to W Kent and northwards to Leicester, but distribution in England very patchy. Recorded from Wales (Merioneth) and Ireland (Co. Armagh; occurrences in Cos. Clare and Down require confirmation). Not reported from Scotland or Isle of Man. Widely distributed and not uncommon in Europe, extending to the Caucasus and Siberia; North Africa.*

21 Smaller, 2.2-3.3 mm; body narrower; tarsi and antennae brown to black, antennal club darker; rostrum generally less robust (pronotum and elytra fig. 327; habitus fig. 328) .... *lutulentus*

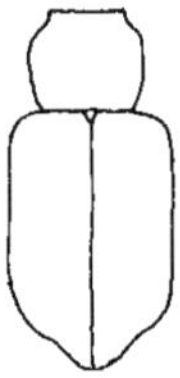

Fig. 327

Differences between sexes small. Male rostrum a little shorter, antennae inserted nearer apex, compared with female rostrum.

In ponds, at the sides of lakes, in ditches, dykes and standing water-bodies generally. Associated with Equisetum fluviatile *(= limosum). In Germany, larvae have been observed to feed in the stems and pupate in their upper parts. Local and not generally common, but widely, if sporadically, distributed in England and Wales from N Somerset eastwards to Surrey and northwards to Anglesey and Cumberland. Also in the Isle of Man and Ireland (S. Co. Kerry, Co. Armagh and Co. Antrim), but not reported from Scotland. Local, but not very rare, throughout Europe, eastwards to Siberia.*

- Larger, 3.5-5.0 mm; body broader and more robust; tarsi, antennal scape (especially distad) and flagellum darker to blackish-brown; rostrum more robust. .... *robustus*

No information on secondary sexual differences available.

Possibly in similar places to those frequented by B. lutulentus, *of which it was once regarded as a form or subspecies, but information scanty. Associated with* Alisma plantago-aquatica *in continental Europe, but biology not known. Extremely rare in the British Isles and probably extinct. A specimen without data, the holotype of* B. rudis *Sharp, is in the Museum of Zoology, Cambridge, and another, a female collected from Hammersmith Marshes, Middlesex, by Sharp, 22 October 1863, is in the National Collection at The Natural History Museum; these are the only British examples known. Widespread in Europe, particularly in the south and recorded from N. Africa.*

112

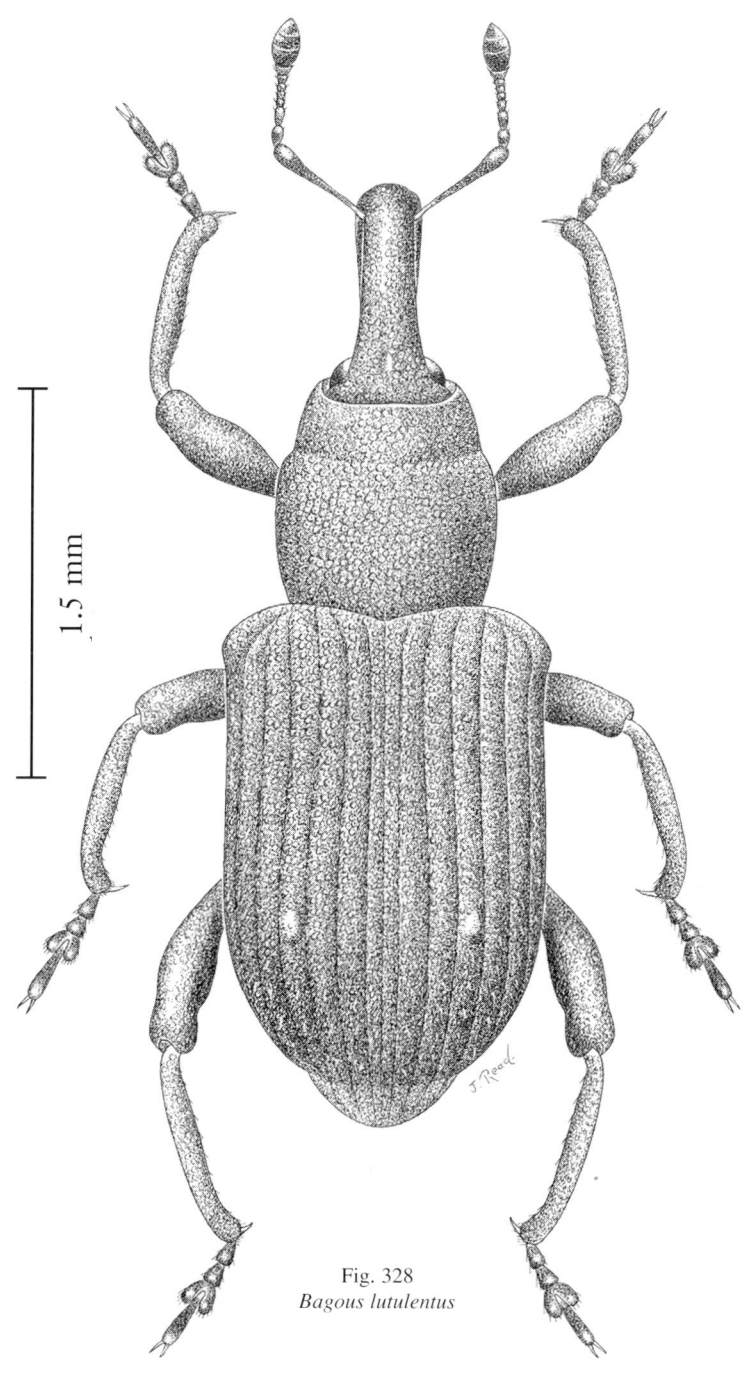

1.5 mm

Fig. 328
Bagous lutulentus

113

Subfamily Dorytominae

The recognition that the subfamily Erirhininae (Erirhinae) of Pope (1977) and other authors is polyphyletic has resulted in the establishment of an 'orthocerous' family Erirhinidae (Thompson, 1992), as previously stated. This has created considerable uncertainty in deciding how to deal with the 'rump' of genera remaining in the old subfamily. Removal of *Bagous* (into Bagoinae) and *Smicronyx* (into Smicronychinae) leaves a disparate and probably polyphyletic group of genera. The solution adopted here is to include subfamilies Styphlinae (for *Pseudostyphlus* and *Orthochaetes*) and Storeinae (for *Pachytychius*), in the knowledge that it is desirable for some amalgamation of subfamilies to take place when their relationships are better understood. Similar problems are raised by the genus *Dorytomus*. Its placement in Curculioninae (*s. l.*) (Zherikhin & Gratshev, 1995; Lawrence & Newton, 1995) would disrupt the arrangement selected for the current series of *Handbooks*, so the least unsatisfactory arrangement appears to assign it to a subfamily Dorytominae, recognising that its correct status is probably only that of a subtribe within Curculioninae-Ellescini (Alonso-Zarazaga & Lyal, 1999). The group contains the single genus *Dorytomus* in the British and European faunas.

Genus *Dorytomus* Germar

About 80 species of *Dorytomus* are known; they occur predominantly in the northern Holarctic, but there are also two South African species. Both O'Brien (1970), who revised the North American species, and Dieckmann (1986), dealing with the fauna of eastern Germany, concluded that the subgenera used in other faunal accounts, for instance Hoffmann (1958), Smreczynski (1972) and Lohse (1983), could not be sustained. These subgenera have not been current in the British literature and they are not used here.

All the species of *Dorytomus* whose biology is known, and certainly the British species, are associated with the genera *Populus* and *Salix* (Salicaceae). The larvae feed in the catkins, although those of some species have been recorded in vegetative buds. Adult weevils frequently overwinter under bark, or in dead wood, of the host trees.

Fourteen species occur in the British Isles, of about twice that number known from central Europe. Despite their small number, the species are among the most difficult of British weevils to identify correctly. Colouration is very variable, some characters are comparative, and care needs to be taken in identifying individuals when voucher material of related species is unavailable. For this reason underside characters have been used more than has been usual in these *Handbooks*.

Key to species

1 Sides of pronotum anteriorly with long, fine setae which conspicuously overlap onto the posterior part of the head (fig. 329). **2**

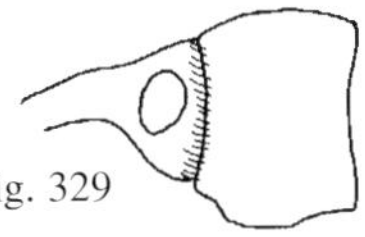

Fig. 329

\- Sides of pronotum with shorter setae which do not overlap onto the posterior part of the head, or do so only slightly and inconspicuously (fig. 330), or without setae. **8**

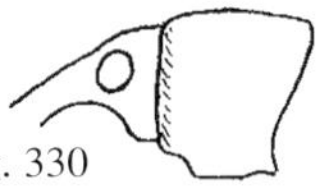

Fig. 330

2 Elytra without upstanding setae, with recumbent pubescence only (fig. 331). **3**

\- Elytra with short, coarse, light and dark, upstanding setae (most easily seen in lateral or oblique view and at elytral apex) as well as recumbent pubescence (fig. 332) [small species 2.8- 3.8 mm]. .... ***hirtipennis***

Fig. 331    Fig. 332

Differences between the sexes not marked. Male rostrum slightly shorter and less strongly curved, antennae inserted at about one-quarter from apex. Female rostrum a little longer and more strongly curved, antennae inserted about one-third from apex.

On river banks, in fens and marshes, in or near wetlands generally. On Salix alba, *recorded in continental Europe less frequently from other narrow-leaved* Salix *species (S. viminalis and S. fragilis, which are native to Britain, and* S. eleagnos, *which has been introduced (Meikle, 1984)). Larvae in both male and female catkins, feeding on stamens and carpels and occasionally boring into the central catkin stem (Morris, 1970). Very local and generally rare. Recorded principally from the Fen country and East Anglia more generally, but with scattered records from Surrey and Berkshire northwards to Chester and S Lancaster. Not reported from western England, Wales or Scotland. Irish records are almost certainly erroneous. Widely distributed throughout northern and central Europe to Siberia, Anatolia and Kasakhstan.*

3 Rostrum long and slender, longer than the head and pronotum combined, often more shining and not, or less evidently, with clothing of scales at base. **4**

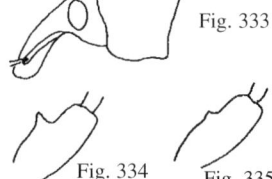

Fig. 333

- Rostrum short and thick, shorter than head and pronotum combined, about as long as pronotum only (fig. 333), dull, and thickly clothed with scales at base [3.9-4.5 mm]. ..... ***ictor***

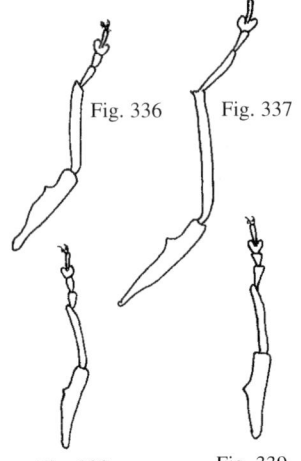

Fig. 334   Fig. 335

Male rostrum slightly shorter and duller at apex; fore-femora much more broadly and strongly toothed (fig. 334). Female rostrum a little longer and more shining at apex; fore-femora with a much smaller tooth (fig. 335).

On river banks, in fields and grazing land and in the dryer parts of fens. On Black Poplar in the wide sense (Populus nigra and its hybrids), but range of cultivars and 'hybrids of inscrutable complexity' (Meikle, 1984) serving as hosts not known in detail. Larvae in catkins. Local and not widely distributed; in south-eastern and central southern England to Suffolk and sporadically northwards to Durham. Not reliably recorded from the south coast or south-west England, but occurring in Wales (Monmouth and Radnor). No records from Scotland, Isle of Man or Ireland. Widely distributed in central Europe (but not extending further north than southern Sweden) and eastwards to eastern Siberia and Japan.

4 Fore-legs not, or not appreciably, longer than mid- or hind-legs (fore-tarsus, fore-tibia and fore-femur each individually not longer than corresponding mid- or hind-segment). **5**

Fig. 336   Fig. 337

- Fore-legs longer (female; fig. 336) or very much longer (male; fig. 337) than mid-legs (female; fig. 338, male; fig. 339) or hind-legs (fore-tarsus, fore-tibia and fore-femur each longer than corresponding mid- or hind-segment) [4.2-6.5 mm] .... ***longimanus***

Male fore-legs exceptionally long, as long as body (excluding rostrum), and thin in proportion (fig. 337); rostrum less strongly curved; antennae inserted nearer rostral apex, apical part of rostrum only half length of antennal scape. Female fore-legs shorter, much less than length of body (excluding rostrum) and more robust (fig. 336); rostrum more stongly curved; antennae inserted further from rostral apex; apical part of rostrum longer, about 0.8 x length of antennal scape.

Fig. 338   Fig. 339

On roadsides, in poplar plantations, grazing land and in the dryer parts of fens and other wetlands. On Populus nigra *and its hybrids, especially* P. x canadensis var. serotina. *Larvae in the catkins. Generally common and widely distributed throughout England and Wales and extending in Scotland as far north as Edinburgh and Stirling. Not recorded from Isle of Man or Ireland. Throughout central and much of southern Europe and as far north as southern Sweden, extending to Siberia and Mongolia; North Africa.*

5 Rostrum strongly to moderately curved, not deflexed at base, forming a distinct angle with dorsum of head in side view; upper margin of eye higher than upper edge of rostrum at base (also in side view) (fig. 340); antennae shorter, neither flagellum nor scape longer than fore-tibia; pronotum more strongly rounded at sides............................. **6**

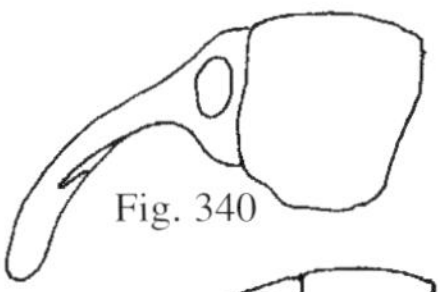

Fig. 340

- Rostrum very slightly curved, almost straight, deflexed at base, forming a smooth outline with dorsum of head in side view; upper edge of rostrum at base higher than upper margin of eye (also in side view) (fig. 341); antennae very long and slender (figs 342, 343), flagellum and scape each appreciably longer than fore-tibia; pronotum much less rounded at sides (fig. 344) [4.0 -5.0 mm]. .... *filirostris*

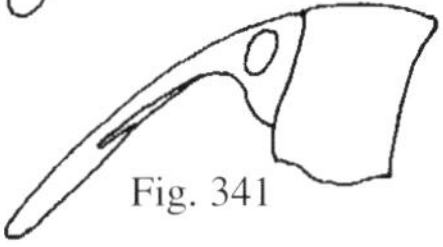

Fig. 341

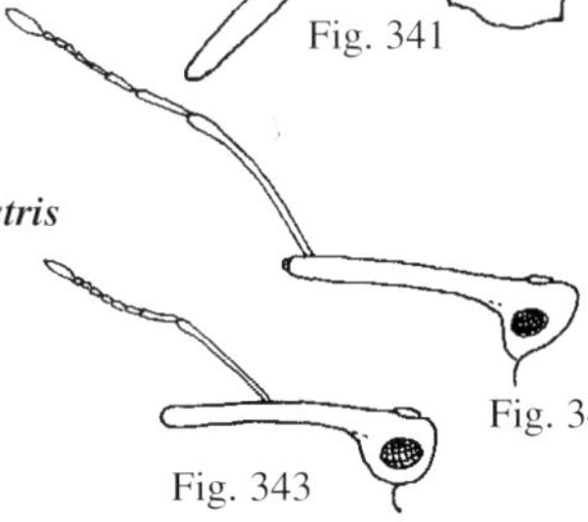

Male antennae longer, scape much longer than width of pronotum; antennal insertion at about 1/6 or less from rostral apex (fig. 342); rostrum slightly more curved, duller and finely carinate; tooth of fore-femur stronger. Female antennae shorter, scape about as long as width of pronotum; antennal insertion at about 1/3 from rostral apex (fig. 343); rostrum slightly straighter, more shining, longitudinal carinae less marked.

Fig. 342

Fig. 343

Fig. 344

In fens and grazing land, at roadsides and in poplar plantations. On Populus nigra *and its hybrids, particularly* P. x canadensis *var.* serotina. *It is also known from* P. alba *in central Europe. The suggestion that it may also feed on* Salix *(Allen, 1947; Donisthorpe, 1947) has not been substantiated and is at variance with experience elsewhere in Europe. Larvae in catkins. Very local and generally scarce, perhaps a relatively recent colonist in Britain, first recorded in 1945 (Allen, 1947). Scattered records, mainly in eastern England, from S Essex northwards to SW and Mid-W Yorks. Not recorded from Wales, Scotland, Isle of Man or Ireland. Throughout most of central Europe, but absent from much of the north, just extending into south-east Denmark and Estonia.*

6 Pronotum smaller, moderately rounded at sides and evidently narrower than elytra at base (fig. 345); rostrum red; elytra more often unicolorous, or bicoloured, less evidently mottled, elytral pubescence sparser; elytral striae less strong; fore-tibia not, or scarcely, angled in middle. **7**

Fig. 345

- Pronotum larger, very strongly rounded at sides, and as broad as elytra at base, or very nearly so (fig. 346); rostrum black; elytra strongly mottled, light and dark, pubescence thicker and more obvious; elytral striae more strongly marked; fore-tibia evidently angled in middle (figs 347, 348) [4.0-5.4 mm]... *tremulae*

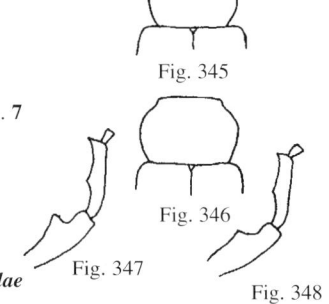

Fig. 346

Fig. 347

Fig. 348

Male rostrum with a longitudinal carina, slightly duller; antennae inserted at about 1/3 from apex, apical portion of rostrum shorter than antennal scape; tooth of fore-femur stronger and broader (fig. 347). Female rostrum without, or with a much less obvious carina, more strongly shining; antennae inserted a little in front of middle of rostrum, apical portion longer than antennal scape; tooth of fore-femur weaker and narrower (fig. 348).

On roadsides, grazing land, waste places and at the edges of woodland. On Populus alba, P. tremula *and, less certainly,* P. x canescens; *also recorded from* P. nigra *in continental Europe. Larvae in vegetative buds and shoots, at least facultatively, but recorded only from catkins in continental Europe. Local and scarce, but widely distributed in England from W Cornwall to Cumberland, though absent from many areas, particularly in the extreme south. Recorded from Wales (Glamorgan); recently discovered in Moray, but not known elsewhere in Scotland. Not recorded from Ireland or Isle of Man. Widely distributed throughout central and northern Europe, extending to Siberia and Asia Minor.*

7 Rostrum less strongly curved (fig. 349); prosternum without an anterior emargination (fig. 350); larger, 4.1-5.3 mm; less frequently bicoloured, usually unicolorous dark red to light orange-red, pubescence very sparse and fine; on *Populus tremula*. .... ***tortrix***

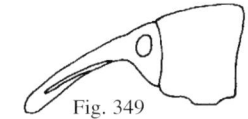

Fig. 349

Fig. 350

Male antennae inserted within anterior third of rostrum; apical portion of rostrum about half as long as antennal scape; rostrum duller, with a more prominent longitudinal carina; tooth of fore-femur larger and stronger. Female antennae inserted within basal two-thirds of rostrum; apical portion of rostrum much longer than half of antennal scape; rostrum more shining, with longitudinal carina less marked; tooth of fore-femur smaller and weaker.

In woods, copses and scrubland, at roadsides and field margins and at streamsides, particularly in northern Britain. On Populus tremula *and not recorded from other* Populus *species either here or abroad. Larvae in catkins, exclusively or predominantly male catkins, at least in Britain (oviposition in late autumn) (Morris, 1998). Generally common and widely distributed throughout England. Fewer records from Wales (Monmouth and Cardigan), and Scotland, but extending to Moray. Local, but widely distributed in Ireland, but not recorded from Isle of Man. Throughout Europe, extending to the Caucasus.*

- Rostrum more strongly curved (fig. 351); prosternum with a conspicuous, deep, anterior emargination (fig. 352); smaller, 3.1-4.0 mm; usually unicolourous red to yellowish but not infrequently bicoloured with elytral disc darker than sides; pubescence thicker and more obvious, particularly at sides of elytra; on species of *Salix*. .... ***melanophthalmus***

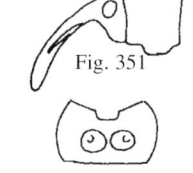

Fig. 351

Fig. 352

Male rostrum shorter, about as long as pronotum and head together, less strongly curved, slightly thicker, less strongly shining, with an evident longitudinal carina; antennae inserted at about one-third from apex; tooth of fore-tibia slightly stronger. Female rostrum longer, evidently longer than pronotum and head together, more strongly curved, a little thinner, more strongly shining, longitudinal carina less evident, evanescent; antennae inserted just in front of middle; tooth of fore-tibia a little weaker.

In woodlands, copses and scrubland, the dryer parts of fens, and at roadsides, in parks and at the sides of lakes, ponds, streams and other water-bodies. On a wide variety of Salix *spp., including* S. caprea, S. cinerea, S. repens, S. aurita *and probably many of their hybrids, but range of hosts not well-known in the British Isles. Larvae in catkins; in Germany both male and female catkins are attacked and oviposition is in late autumn, the eggs overwintering, but the biology has not been studied in Britain. Somewhat local but quite common and widely distributed throughout England and Wales, though not recorded from the south-west Peninsula. Few records from Scotland, but extending to Moray. Recorded from Ireland (N Co. Kerry) but not from Isle of Man. Widely distributed throughout Europe and extending to the Caucasus and North Africa.*

8 Prosternum deeply emarginate anteriorly, the emargination angled at sides (fig. 353). **12**

Fig. 353

- Prosternum not emarginate, or with a shallow, weak emargination which is not angled at sides (cf. fig. 352). **9**

9 Rostrum shorter than, or at most as long as, head and pronotum together (fig. 354); elytra shorter, slightly more rounded at sides and less depressed on disc, 1.55-1.67 x as long as broad; subapical prominence with patch of pale pubescence on elytral interstice 5 inconspicuous or absent [rare and uncommon species]... **10**

Fig. 354

117

- Rostrum longer than head and pronotum together (fig. 355); elytra longer, less rounded at sides (almost parallel-sided or sides divergent apicad) and more depressed on disc, 1.69-1.75 x as long as broad; subapical prominence with patch of pale pubescence on elytral interstice 5 generally conspicuous [common and widely distributed species]... **11**

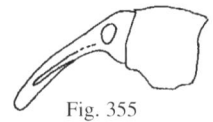

Fig. 355

10 Pronotum evenly rounded at sides, less abruptly narrowed at apex, broadest in middle, not, or scarcely, broader at base than at apex (fig. 356); elytra less elongate, 1.54-1.57 x as long as broad; smaller insect, 2.5-3.1 mm; on *Salix repens* [generally paler, orange-brown; recently recorded only from Yorkshire]. .... *salicis*

Fig. 356

Male rostrum slightly shorter, thicker and less shining, especially apicad; antennae inserted nearer apex, about 1/4 from tip of rostrum, its apical portion about half length of antennal scape and scarcely longer than broad at widest point; tooth of fore-tibia broader and stronger. Female rostrum slightly longer, thinner and more shining towards apex; antennae inserted further from apex, at about 1/3 from apex of rostrum or a little less, apical portion of rostrum clearly more than half the length of antennal scape and evidently longer than broad at widest point; tooth of fore-tibia narrower and weaker.

On moors, fens, commons and roadsides. Associated principally with Salix repens *in Britain, but possibly on other species* (S. caprea, S. cinerea, S. aurita *and* S. alba), *as in central Europe. Biology unknown, either in the British Isles or Europe, but larvae probably in catkins. Very local and generally rare. Recorded from a very few southern counties of England, but not recently; it is possible that some of these records are unreliable. Known from a small number of sites in SE, NE and SW Yorkshire, with older records from N Northumberland. Not reported from Wales, Scotland, Ireland or Isle of Man. Throughout central and northern Europe but everywhere scarce and uncommon.*

- Pronotum not evenly rounded at sides, very abruptly narrowed at apex, often broadest in front of middle, much broader at base than at apex (fig. 357); elytra more elongate, 1.63-1.67 x as long as broad; larger insect, 3.6 - 4.2 mm; on *Populus tremula* [generally darker, blackish-brown; recently recorded only from E Kent and Huntingdon]. .... *affinis*

Fig. 357

Differences between sexes not marked. Male rostrum slightly shorter, antennae inserted nearer to apex, apical portion of rostrum about as long as broad at insertion; tooth of fore-tibia a little broader at base. Female rostrum a little longer, antennae inserted rather further from apex, apical portion of rostrum distinctly longer than broad at insertion; tooth of fore-tibia less broad at base.

In woods, woodland glades and at ride margins. On Populus tremula *and not recorded from other species of* Populus *in Britain, although an adventitious association with* Quercus *has been noted by some observers (Allen, 1968). In central Europe it has been recorded less commonly on* Populus nigra. *Larvae in female catkins only (Morris, 1998). Very local and rare, but often in numbers where found. Known only from E and W Kent, Cambridge and Huntingdon, a Leicester record requiring confirmation. Other, older, records mostly erroneous (Allen, 1967). Central and northern Europe to Siberia. RDB2.*

11 Rostrum more strongly and regularly curved, particularly between base and antennal insertion (fig. 358); antennae longer, segments 5-7 of funiculus quadrate or at most very slightly transverse, club slightly longer (fig. 359); colour of fore-tarsus usually much lighter than tibial base (tibia often quite strongly bicolorous); insect larger 3.7-5.0 mm; on *Populus tremula*, less often on other species of *Populus*. .... *dejeani*

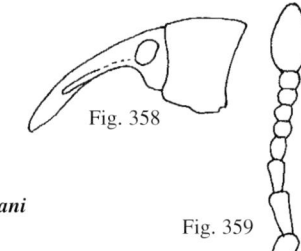

Fig. 358

Fig. 359

Male rostrum shorter, only just longer than head and pronotum together; antennae inserted nearer apex, apical portion of rostrum about half as long as antennal scape; apical tooth of fore-tibia longer and stronger (fig. 360). Female rostrum longer, much longer than head and pronotum together; antennae inserted further from rostral apex, apical portion of rostrum longer than half length of antennal scape; apical tooth of fore-tibia smaller and less well developed (fig. 361).

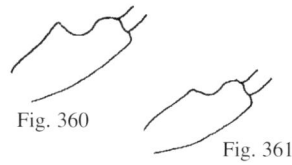

Fig. 360

Fig. 361

In woods, woodland glades and at the sides of rides, at roadsides and wherever the host plant grows. On Populus tremula *and not, or seldom, on any other species of* Populus *in the British Isles, though recorded from* P. alba *and* P. nigra *in continental Europe. Larvae in catkins, mainly, though not exclusively, female catkins (oviposition in early spring) (Morris, 1998). Fairly common and widely distributed throughout England; less common in Wales (Cardigan) and Scotland, though extending as far north as Moray. Not recorded from Ireland or Isle of Man. Widely distributed throughout Europe and extending to the Caucasus.*

-   Rostrum less strongly and less regularly curved, basal portion to antennal insertion almost straight (fig. 362); antennae shorter, segments 5-7 of funiculus evidently transverse, club slightly shorter (fig. 363); fore-tarsus and base of tibia usually unicolorous (tibia seldom bicolorous); insect smaller 3.0-4.5 mm; on species of *Salix*.
    .... ***taeniatus***

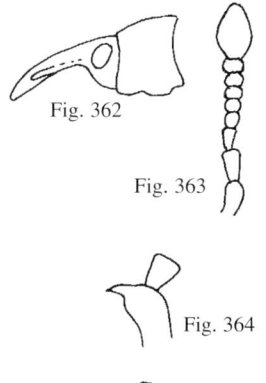

Fig. 362

Fig. 363

Male rostrum only as long as, or shorter than, head and pronotum together; antennae inserted nearer rostral apex, apical portion of rostrum shorter than half length of antennal scape; tooth of fore-femur slightly stronger and broader; apical tooth of fore-tibia also a little longer and more strongly developed (fig. 364). Female rostrum evidently longer than head and pronotum together; antennae inserted a little further from rostral apex, apical portion of rostrum clearly longer than half length of antennal scape; tooth of fore-femur a little weaker and less broad; apical tooth of fore-tibia slightly shorter and less well-developed (fig. 365).

Fig. 364

Fig. 365

In woods, copses, scrubland, fens, moors and wetlands; at roadsides, along water courses and in hedgerows, parks and grazing land. On a variety of broad-leaved Salix *species, including* S. capraea, S. cinerea, S. aurita *and probably many of their hybrids. Records from* Populus *spp. require confirmation and may be due to confusion with other species or casual occurrence. Larvae in catkins, feeding on pistils and ovules (and doubtless stamens in male catkins) and burrowing into the central catkin stems (Cawthra, 1957b). Occasionally reported as feeding in vegetative shoots, for example of* S. reticulata *(Bland, 1997). Common and widespread throughout the British Isles, by far the commonest species of the genus. Recorded from most of the vice-counties of England and Wales and also widespread in Ireland and Scotland, though not recorded from the Hebrides, Orkneys or Shetland. Also not reported from the Isle of Man. Common throughout Europe to the Caucasus and Siberia. North Africa.*

12 Pronotum strongly transverse, about 1.25 x as broad as long (cf. fig. 366); elytra shorter and more rounded at sides, 1.50-1.66 x as long as broad, less depressed on disc.
.... **13**

Fig. 366

-   Pronotum slightly elongate, or quadrate to slightly transverse, but never more than 1.1 x as broad as long (fig. 367); elytra longer and less rounded at sides, 1.79-1.82 x as long as broad, more depressed on disc.................. ***salicinus***

Fig. 367

Male rostrum shorter, a little shorter than head and pronotum together; antennae inserted nearer apex, within apical third; apical portion of rostrum little more than half length of antennal scape; tooth of fore-femur a little longer and broader (fig. 368). Female rostrum longer, longer than head and pronotum together; antennae inserted further from rostral apex, at about one-third of its length from apex; apical portion of rostrum about 0.75 x length of antennal scape; tooth of fore-femur slightly shorter and narrower (fig. 369).

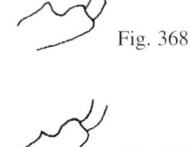

Fig. 368

Fig. 369

In fens and other wetlands, often in wetter places than those in which other sallow-feeding species are found. On Salix *species, probably mainly* S. cinerea, *but also* S. caprea *and* S. aurita. *Biology little-studied either in Britain or continental Europe; larvae probably in catkins. Very local, mainly restricted to E. Anglia, the East Midlands and Northern England, but also recorded from Dumfries though not known elsewhere in Scotland. Records from N Wiltshire and Glamorgan require confirmation, but the species has been found very recently in Dorset. Not recorded from Isle of Man or Ireland. Northern and central Europe to Siberia.*

13 Larger species, 3.1-4.0 mm; rostrum not rugosely punctured, somewhat smooth and shining, punctures discrete; elytra with very sparse pubescence, mainly, or only, at sides and at apex, not obscuring underlying sculpture or striae; [common and widely distributed species, frequent in southern England]. **14**

- Smaller species, 2.4-3.0 mm; rostrum rugosely punctured, dull, punctures confluent; elytra (also pronotum and rostral base) more thickly covered with pubescence, especially at sides; [rare and seldom recorded; northern England and southern Scotland only]... ***majalis***

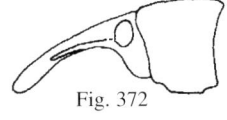

Fig. 370

Fig. 371

Male rostrum shorter, shorter than, or about as long as, head and pronotum together; antennae inserted nearer rostral apex, about one-third from apex; tooth of fore-femur stronger and broader (fig. 370). Female rostrum longer, longer than head and pronotum together; antennae inserted further from rostral apex, more than one-third from apex; tooth of fore-tibia weaker and finer (fig. 371).

In woods, copses and scrubland, by water courses and in fens and bogs. On Salix cinerea, S. aurita *and* S. caprea, *though recorded from a variety of other* Salix *species in central Europe. Larvae in catkins, but biology little-studied in Britain. Rare and seldom recorded; no records in the last 40 years. Reported from a very few vice-counties in northern England and southern Scotland; last record from Cumberland pre-1960. As with several other species of* Dorytomus, *some doubt exists about the correctness of some earlier records. Widely distributed in northern and central Europe. Proposed RDB (insufficiently known) (Hyman & Parsons, 1992).*

14 Rostrum less strongly and regularly curved, almost straight to insertion of antennae, weakly curved apicad (fig. 372), shorter, in male shorter than head and pronotum together, in female about as long as head and pronotum together; pronotum less regularly curved at sides, less constricted basad and broadest in front of middle (fig. 373) [3.3-4.0 mm] (habitus fig. 374).....................................***rufatus***

Fig. 372

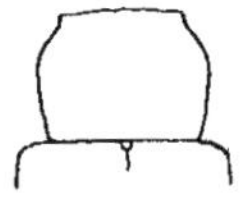

Fig. 373

Male rostrum shorter, shorter than head and pronotum together, more robust and duller, strongly punctured on either side of median carina; antennae inserted nearer rostral apex, apical portion of rostrum about 0.6 x length of antennal scape; tooth of fore-femur broader and stronger. Female rostrum longer, about as long as head and pronotum together, thinner and more strongly shining, less strongly punctured on either side of median carina; antennae inserted further from rostral apex, apical portion of rostrum about 0.75 x length of antennal scape; tooth of fore-femur weaker, narrower and more sharply pointed.

In woods, copses, scrubland, fens, moors and bogs and at roadsides, in parks and by watercourses.

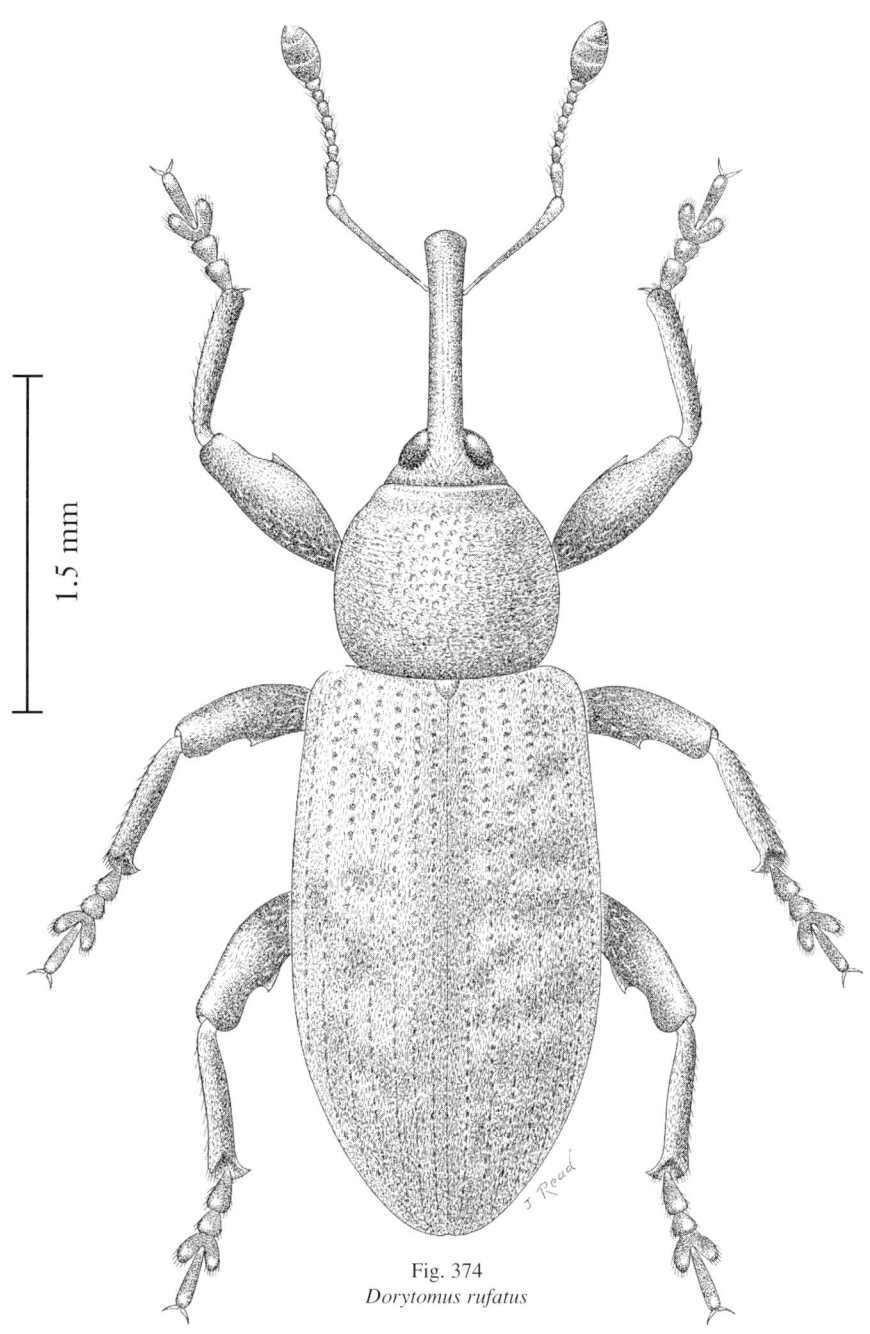

1.5 mm

Fig. 374
Dorytomus rufatus

On broad-leaved species of Salix, *especially* S. caprea *and* S. aurita, *but range of host plants not well-known in Britain. Larvae in catkins, probably male and female. Fairly common and widely distributed throughout England, Wales, Ireland and Scotland except the far north (most northerly record from Easterness); Isle of Man. Common and widely distributed throughout Europe to the Caucasus.*

- Rostrum strongly and regularly curved (fig. 375), longer, in both sexes longer than head and pronotum together; prothorax more regularly curved at sides, more constricted basad, and broadest at middle (fig. 376) (see couplet 7).
.... ***melanophthalmus***

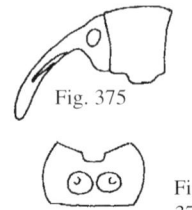

Fig. 375

Fig. 376

Subfamily Storeinae

Pachytychius is an enigmatic genus and the relationship of it and its allies to other groups of curculionine weevils (in the broad sense) is uncertain. The genus has been placed usually in Erirhininae (= Notarinae) in the old sense (Pope, 1977; Caldara, 1978; Lohse, 1983; Abbazzi & Osella, 1992), but the recognition that this group is polyphyletic and the exclusion of a number of genera formerly placed therein from Erirhinidae (*sensu* Thompson, 1992) leaves *Pachytychius* isolated. Provisional positioning in a subfamily Storeinae appears to be the least unsatisfactory solution. The group is placed in Curculioninae-Storeini by Alonso-Zarazaga & Lyal (1999).

Genus *Pachytychius* Jekel

About 40 species of *Pachytychius* are known from the Palaearctic; the fauna was revised by Caldara (1978). The Mediterranean region is richer in species than northern and central Europe, into which area only two species extend. Of these only one inhabits the British Isles.

- One species; small (3.0-3.9 mm), oval-elongate; pronotum as broad as elytra at base (fig. 377); rostrum strongly curved; elytra, and posterior half of pronotum at sides, thickly covered with appressed variegated scales; hind femora with a strong, sharp tooth, fore- and mid-femora unarmed.. ***haematocephalus***

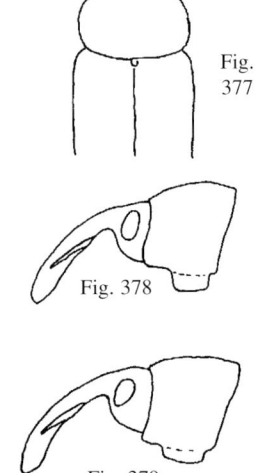

Fig. 377

Fig. 378

Fig. 379

Male rostrum duller, slightly shorter, shorter than head and pronotum together, less strongly curved (fig. 378), median keel stronger; antennae inserted in apical third of rostrum. Female rostrum more shining, longer, about as long as head and pronotum together, more strongly curved (fig. 379), median keel less well-developed; antennae inserted behind apical third of rostrum.

In coastal grassland, possibly other grasslands, and open country where the foodplant grows; probably xerothermic, occurring only on south-facing slopes. On Lotus corniculatus *(and other* Lotus *spp. on the Continent). Larvae feed on seeds in the ripening pods. Rare and very local. It persists at one site near Gosport, S Hants, and has been recorded from Dorset and N Wilts.; the latter record is 120 years old and requires confirmation. No Welsh, Scottish or Irish records. A Mediterranean species on the edge of its range in England. Central, western, southern and south-eastern Europe to Asia Minor and western North Africa. RDB1.*

122

Subfamily Styphlinae

This is another group that was included in Erirhininae in the old sense, but which requires a name following the division of that group into its 'orthocerous' (erirhinid) and 'gonatocerous' constituents. The group is placed as a tribe, Styphlini, of Curculioninae by Alonso-Zarazaga & Lyal (1999).

Key to genera

1   Scutellum clearly evident (fig. 380); antennal funiculus with 7 segments (fig. 381); shoulders of elytra well developed (fig. 380); pronotum and elytra thickly clothed with broad, isodiametric, closely appressed, pale scales (elytra with erect setae also), though often with an encrustation of soil; pronotum transverse, about 1.2 x as broad as long, sides rounded, weakly sinuate and strongly convergent anteriad; raised interstices of elytra less strongly developed, but third interstice evidently raised at base, erect elytral setae smaller; elytra broadest in front of middle; rostrum without a fovea, or transverse impression, at base; apex of antennal scape with four or five erect, conspicuous, setae (sometimes eroded). .... ***Pseudostyphlus***

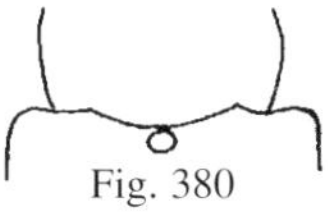

Fig. 380

Fig. 381

-   Scutellum wanting (fig. 382); antennal funiculus with 6 segments (fig. 383); shoulders of elytra effaced (fig. 382); pronotum and elytra without a clothing of appressed scales; pronotum quadrate to very weakly transverse, about 1.0-1.1 x as broad as long, slightly rounded at sides, weakly convergent to subparallel anteriad; raised interstices of elytra more strongly developed, especially at base, erect setae larger and more evident; elytra broadest at middle; apex of antennal scape with only one or two conspicuous, erect setae (sometimes eroded). .... ***Orthochaetes***

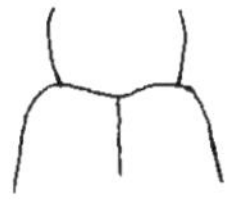

Fig. 382

Fig. 383

Genus *Pseudostyphlus* Tournier

Only three species are known in this genus in the Palaearctic region and, of these, just one in western and central Europe. The misspelling of the trivial name *pillumus*, widespread in the older literature, is fully discussed by Dieckmann (1986).

-   One species; small (2.5-3.6 mm), oblong-oval, slightly depressed; antennae, legs and apex of rostrum brown to red-brown; femora and tibiae with broad, apressed, isodiametric scales and raised setae; rostrum with a median longitudinal keel and less well-developed keels at sides; other characters as in generic key. .... ***pillumus***

Fig. 384

Male rostrum slightly shorter, shorter than head and pronotum together; antennae inserted in anterior third of rostrum (fig. 384). Female rostrum a little longer, as long as or slightly longer than head and pronotum together; antennae inserted a little further from rostral apex (fig. 385).

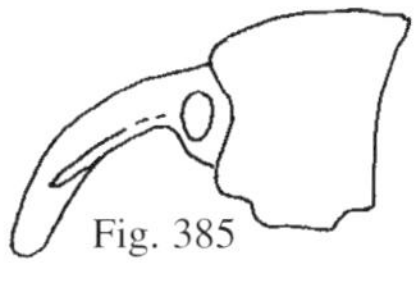

Fig. 385

In waste places and disturbed land, at field margins, and in open coastal areas, most often in arenaceous places. On a variety of Asteraceae-Anthemideae in continental Europe, including species of Matricaria, Tripleurospermum, Anthemis *and* Achillea. *Hosts less well-known in Britain, but most usually associated with*

Matricaria recutita. Larvae in capitula, many to one head, though biology little-studied in the British Isles. Fully winged and flightless individuals are known from Germany, but the phenomenon of alary polymorphism in P. pillumus *has not been studied in Britain. Local and generally rare, though often in numbers where found. In southern and south-east England only, recorded from S Hants eastwards to E Kent and northwards to Buckingham and Cambridge. Not recorded from south-western, western or northern England, nor from Wales, Scotland or Ireland. Widely distributed in Europe northwards to southern Fennoscandia.*

Genus *Orthochaetes* Germar

Gonzalez (1967) revised the species of Orthochaetes occurring in western Europe, but his work is largely unknown to British coleopterists. He included 21 species in his account, but this total includes eight in the non-British subgenera *Styphlus, Styphlidius* and *Trachysoma* (=*Trachystyphlus*), which other workers (e.g. Dieckmann, 1986; Alonso-Zarazaga & Lyal, 1999) have regarded as good genera. *Orthochaetes* are small, rather obscure, ground-living weevils, the larvae of which are leaf-miners on herbs. Two species are known to occur in the British Isles.

Key to species

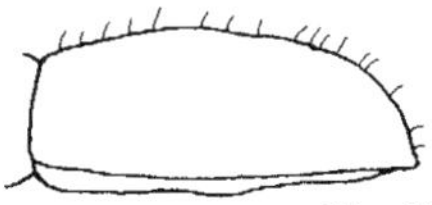

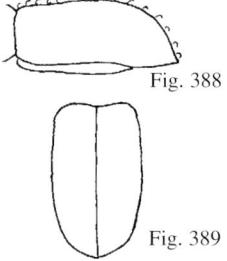

Fig. 386

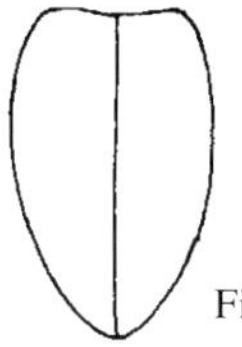

Fig. 387

1 Upstanding setae of elytral interstices fully erect, only very slightly curved (fig. 386); elytra ovoid, strongly rounded at sides, more gradually convergent to apex (fig. 387); on average slightly larger, 2.6-3.5 mm (habitus fig. 390).
   .... *setiger*

Parthenogenetic throughout most of Europe, though a male has been taken in Spain; although British specimens have not been examined extensively males have not so far been found.
   In grasslands (common on chalk downland) and in open situations; often amongst moss and under stones. Polyphagous.
Larvae are leaf-miners in a very wide range of Asteraceous herbs, and also in species of other families, in continental Europe. There are few references to feeding in the British literature but a record of leaf-mining Allium ursinum *(Bland & Nelson, 1997) is notable as being from a monocotyledonous plant (cf.* O. insignis*). Flightless and ground-living. Much commoner than the literature suggests, often abundant in vacuum samples from grassland. Widespread throughout England and Wales, but infrequently recorded from southern Scotland, though occurring as far north as Linlithgow. Local, though widely distributed, in Ireland and recorded from Isle of Man. Throughout most of central and northern Europe except the far north and excluding the Mediterranean region.*

- Upstanding setae of elytral interstices strongly curved backwards, apices often almost touching elytra (fig. 388); elytra rectangular, subparallel-sided, more abruptly convergent to apex (fig. 389); on average a little smaller, 2.4-3.0 mm. ....*insignis*

Fig. 388

Fig. 389

Probably parthenogenetic in Britain, though there is no certain confirmation of this. Males are known elsewhere from the Tangiers region of Morocco (Dieckmann, 1986); the illustration of the median lobe of the male in Tempere & Pericart (1989) is taken from Gonzalez (1967).
   In grasslands, coastal undercliffs and shingle banks, in disturbed ground and in ruderal plant communities. Flightless and mainly ground-living, though occasionally found on taller vegetation. Exceptionally polyphagous, probably the most widely polyphagous phanerognathous weevil known. Recorded as mining the leaves of herbs in 14 plant families, including monocotyledous as well as dicotyledonous ones (Hering, 1957). Very local and generally rare; often taken as single specimens. Coastal counties of England from W Cornwall to E Kent, also recorded from Surrey, Wales (Glamorgan, Pembroke and Anglesey) and the Burren region of Co. Clare, but not known from central or northern England, Scotland or the Isle of Man. A narrow European and N. African distribution, in the western part of the area only.

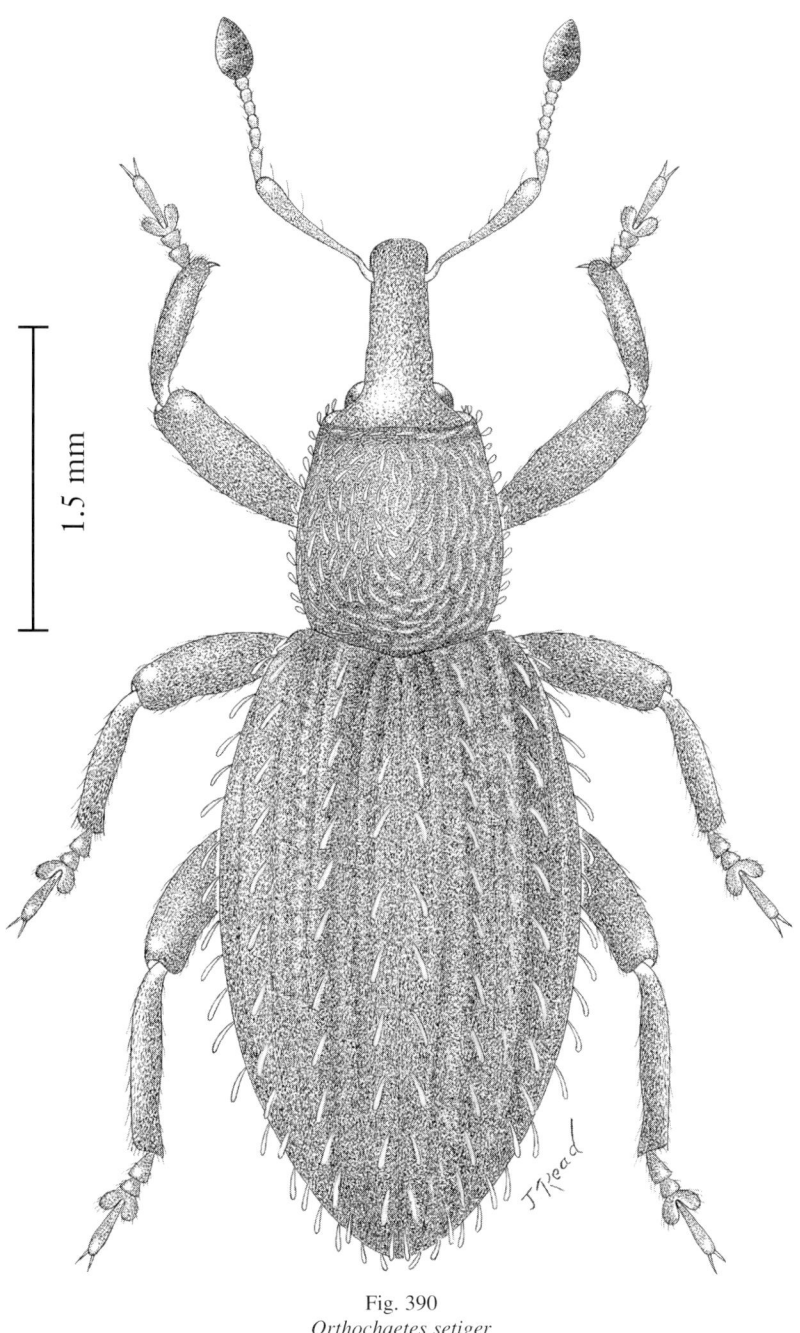

Fig. 390
Orthochaetes setiger

Subfamily Smicronychinae

This group has previously been treated as a tribe of Erirhininae but has been regarded as a subfamily following partial revision of the gonatocerous members of that group (Thompson, 1992). The subfamily is almost cosmopolitan but is not represented in the Ethiopian or Australasian realms; most of the 160 or so species are found in the Palaearctic and Nearctic regions. Six genera are known, of which two occur in the European fauna but only one in the British Isles. Alonso-Zarazaga & Lyal (1999) treat Smicronychini as a tribe of Curculioninae.

Genus *Smicronyx* Schoenherr

About 40 species of *Smicronyx* are known from the Palaearctic region, but the genus is more speciose in North America where about twice that number occur. Ten species inhabit western and central Europe, of which three occur in Britain.

The European species of *Smicronyx* are small (1.6-2.3 mm) weevils, many of which have parasitic or semiparasitic plants, especially dodders (*Cuscuta* spp.) as hosts.

Key to species

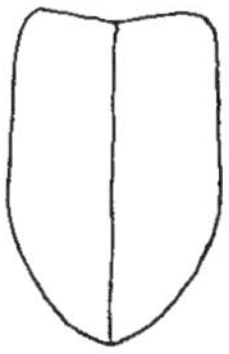

1 Pronotum punctured, punctures shallow, interspaces shining; elytra slightly more elongate, 1.4-1.5 x as long as broad and more nearly parallel-sided (fig. 391); on average slightly smaller, 1.8-2.3 mm [on *Cuscuta*]. **2**

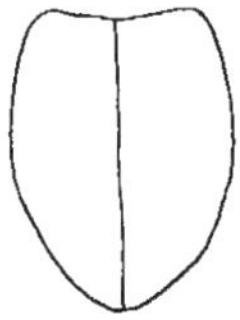

- Pronotum corniculate or finely tuberculate, less strongly shining; elytra less elongate, 1.3-1.4 x as long as broad, more strongly rounded at sides (fig. 392); on average slightly larger, 2.0-2.5 mm [mainly on *Centaurium*]. .... *reichi*

Fig. 391

Fig. 392

Male rostrum in apical half more strongly punctured, dull, in basal half thickly clothed with fine, pale pubescence. Female rostrum in apical half less strongly punctured, more shining, in basal half without, or with much sparser, pubescence, almost glabrous.

On chalk and limestone grassland, on calcareous roadside verges, and in sand- and chalk-pits. On Centaurium erythraea, *possibly other* Centaurium *spp., and, less certainly, on* Blackstonia perfoliata *and* Gentianella *spp. (plants which are not known as hosts in continental Europe, though* Gentianella *spp. are foodplants of the closely-related* S. swertiae *Voss). Larvae in seed-heads, pupating in soil (Blair, 1935a). Local and not common, though less rare than records suggest, in southern England only, from S Devon eastwards to E Kent and northwards to Berkshire and W Gloucester. Not recorded from northern England, Wales, Scotland or Ireland. Central and southern Europe to Anatolia. The proposal for RDB3 status (Hyman & Parsons, 1992) does not appear to be justified; the species is clearly under-recorded in Britain.*

2 Tarsal claws unequal (fig. 393); rostrum more strongly curved (fig. 394); upper surface with only small, fine, pale setae which do not overlap (frequently eroded) [1.8-2.3 mm]. .... *coecus*

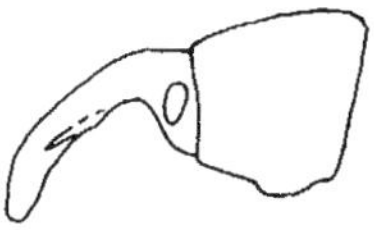

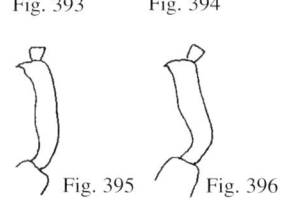

Fig. 393   Fig. 394

Male fore-tibia with a longer and stronger apical tooth (fig. 395); antennae inserted nearer rostral apex, apical portion of rostrum shorter than antennal scape and less strongly shining. Female fore-tibia with a shorter and weaker apical tooth (fig. 396); antennae inserted further from rostral apex, apical portion as long as, or longer, than antennal scape.

On cliffs, shingle beaches, acid grasslands and heaths. On Cuscuta epithymum *and possibly* C. europaea, *as elsewhere.*

Fig. 395   Fig. 396

126

Fowler (1891), who knew only two British examples, associated it only with C. europaea, *but this is a rare and declining species compared with* C. epithymum *(Stace, 1991). The latter species seems to be the usual host in Britain, as far as can be judged from the few records of* S. coecus *here. (C.* epithymum *parasitises* Cytisus scoparius *and* Ulex europaeus, *in particular;* C. europaea *usually parasitises* Urtica dioica). *Larvae in stem-galls on the host. Very local and scarce; usually taken with, but less abundant than,* S. jungermanniae. *Recorded from only four coastal vice-counties of southern England: N Devon, Isle of Wight, Dorset and E Kent. Widespread in Europe as far north as Denmark and Estonia and eastwards to the Caucasus. RDB3.*

- Tarsal claws equal (or nearly so) (fig. 397); rostrum less strongly curved (fig. 398); upper surface with larger, broad, variegated scales (which, however, are often eroded) [1.8-2.3 mm]. .... *jungermanniae*

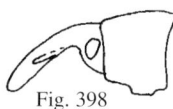

Fig. 397

Male rostrum shorter, more strongly punctured, duller and more pubescent basally; antennae inserted nearer rostral apex, apical portion of rostrum shorter than antennal scape; apical tooth of fore-tibia stronger and longer (cf. fig. 395). Female rostrum slightly longer, less strongly punctured, more strongly shining and less pubescent; antennae inserted further from rostral apex, apical portion of rostrum as long as, or longer than, scape; apical tooth of fore-tibia smaller and weaker (cf. fig. 396).

Fig. 398

On heaths, cliff-tops, acid grasslands and shingle beaches. Chiefly on Cuscuta epithymum *(possibly also on other* Cuscuta *spp., though these are rare in Britain). Larvae in stem-galls. Local, and of uncertain appearance, mainly because of the transient and variable occurrence of its host, but often in numbers when found. In southern England only, coastal vice-counties from W Cornwall to E Kent and northwards to Berkshire and E Norfolk. Not recorded from central and northern England, Wales, Scotland or Ireland. Throughout Europe to central Asia. North Africa. The commonest and most widely distributed of the European species in the genus.*

127

References

ABBAZZI, P. & OSELLA, G., 1992. Elenco sistematico-faunistico degli Anthribidae, Rhinomaceridae, Attelabidae, Apionidae, Brentidae, Curculionidae Italiani (Insecta, Coleoptera, Curculionoidea. *Redia* **75**: 267-414.

ALFORD, D.V., 1984. *A Colour Atlas of Fruit Pests.* Wolfe Science, London.

ALLEN, A.A., 1947. *Dorytomus filirostris* Gyll. (Col., Curculionidae), a weevil new to Britain. *Entomologist's monthly Magazine* **83**: 52.

ALLEN, A.A., 1968. *Dorytomus affinis* Payk. (Col., Curculionidae) in Kent and notes on its British allies. *Entomologist's monthly Magazine* **103** (1967): 264-267.

ALLEN, A.A., 1972a. *Magdalis memnonia* Gyll. (Col., Curculionidae), a weevil new to Britain. *Entomologist's Record and Journal of Variation* **84**: 22-23.

ALLEN, A.A., 1972b. A contribution to the knowledge of *Hypera meles* F. (Col., Curculionidae) in Britain. *Entomologist's Record and Journal of Variation* **84**: 110-113.

ALLEN, A.A., 1976. *Magdalis memnonia* Gyll. (Col., Curculionidae) established as a British-breeding species. *Entomologist's monthly Magazine* **111** (1975): 96.

ALLEN, A.A., 1982. *Magdalis violacea* L. (Col.: Curculionidae): correction of a record. *Entomologist's Record and Journal of Variation* **94**: 120-121.

ALONSO-ZARAZAGA, M.A. & LYAL, C.H.C., 1999. *A world catalogue of families and genera of Curculionoidea (Insecta: Coleoptera) (Excepting Scolytidae and Platypodidae).* Museo Nacional de Ciencias Naturales, Madrid & The Natural History Museum, London.

ANDERSON, R.S., 1995. An evolutionary perspective on diversity in Curculionoidea. *Memoirs of the entomological Society of Washington* **14**: 103-114.

ANDERSON, W.H., 1948. A key to the larvae of some species of *Hypera* Germar, 1817 (= *Phytonomus* Schoenherr, 1823) (Coleoptera, Curculionidae). *Proceedings of the entomological Society of Washington* **50**: 25-34.

ASHE, G.H., 1944. *Hylobius transversovittatus* Goeze (= *fatuus* Rossi) (Col., Curculionidae) new to Britain. *Entomologist's monthly Magazine* **80**: 287.

BARRIOS, H.E., 1986. A review of weevils of the genus *Magdalis* Germar (Coleoptera, Curculionidae) of the fauna of the European part of the USSR and the Caucasus. *Entomological Review* **66**: 25-45.

BEARE, T.H., 1922. *Hypera meles* F. and other Coleoptera in a lucerne field at Wicken. *Entomologist's monthly Magazine* **58**: 249.

BEARE, T.H., 1930. *Pissodes validirostris* Gyll., a British insect. *Entomologist's monthly Magazine* **66**: 274.

BEVAN, D., 1987. *Forest Insects.* (Forest Commission Handbook 1). HMSO, London.

BLAIR, K.G., 1935a. *Smicronyx reichi* Gyll., with notes on other species of the genus. *Entomologist's monthly Magazine* **71**: 127-130.

BLAIR, K.G., 1935b. *Bagous frit* Herbst. in Britain, with notes on some other species of the genus. *Entomologist's monthly Magazine* **71**: 249-253.

BLAIR, K.G., 1948. Some alien Coleoptera occasionally found in Britain. *Entomologist's monthly Magazine* **84**: 123-124.

BLAND, K.P., 1997. Atypical lifestyle for *Dorytomus taeniatus* (Fab.) (Col., Curculionidae) on *Salix reticulata* L. *Entomologist's monthly Magazine* **133**: 172.

BLAND, K.P. & NELSON, J.M., 1997. *Orthochaetes setiger* (Beck) (Coleoptera: Curculionidae) leaf-mining *Allium ursinum* in southern Scotland. *British Journal of Entomology and natural History* **10**: 65-67.

BLOSSEY, B. 1993. Herbivory below ground and biological weed control: life history of a root-boring weevil on purple loosestrife. *Oecologia* **94**: 380-387.

BOOTH, R.G., 1993. *Thryogenes fiorii* Zumpt, 1928 (Curculionidae) new to Britain. *Coleopterist* **2**: 19-20.

BOOTH, R.G., in press

BUCK, F.D., 1948. *Pentarthrum huttoni* Woll. (Col., Curculionidae) and some imported Cossoninae. *Entomologist's monthly Magazine* **84**: 152-154.

BURRINI, A.G., MAGNANO, L., MAGNANO, A.R. & BACCETTI, B., 1988. Spermatozoa and phylogeny of Curculionoidea (Coleoptera). *International Journal of Insect Morphology & Embryology* **17**: 1-50.

CALDARA, R., 1978. Revisione dei *Pachytychius* paleartici (Coleoptera Curculionidae). *Memorie della Società entomologica Italiana* **56** (1977): 131-216.

CALDARA, R. & O'BRIEN, C.W., 1998. Systematics and evolution of weevils of the genus *Bagous* VI. Taxonomic treatment of the species of the western Palaearctic Region (Coleoptera

Curculionidae). *Memorie della Società entomologica Italiana* **76**: 131-347.

CALLAN, E.McC., 1939. Insects bred from stumps of *Salix viminalis* L. *Journal of the Society for British Entomology* **2**: 21-22.

CAWTHRA, E.M., 1957a. Some notes on *Grypidius equiseti* F. (Col., Curculionidae) with a description of its larva. *Proceedings of the Royal entomological Society of London (A)* **32**: 95-106.

CAWTHRA, E.M., 1957b. Notes on the biology of a number of weevils (Col., Curculionidae) occurring in Scotland. *Entomologist's monthly Magazine* **93**: 204-207.

CAWTHRA, E.M., 1958. The occurrence of *Cleonus piger* Scop. (Col., Curculionidae) at Aberlady Bay, East Lothian, with some notes upon its larval instars. *Entomologist's monthly Magazine* **94**: 204-206.

CHAMPION, G.C., 1879. *Bagous diglyptus* Boh. (a species new to the British list) at Burton-on-Trent. *Entomologist's monthly Magazine* **15**: 235.

CLEMONS, L., 1983. *Gronops inaequalis* Boheman (Col.: Curculionidae): a weevil new to Britain. *Entomologist's Record and Journal of Variation* **95**: 213-215.

COLLINGWOOD, C.A., 1954. The biology of *Epipolaeus caliginosus* F. (Col., Curculionidae). *Entomologist's monthly Magazine* **90**: 169-172.

COPESTAKE, D.R., 1999. Biological notes on *Bagous limosus* Gyllenhal (Curculionidae), including locomotion and respiration. *Coleopterist* **8**: 63-67.

COX, L.G., 1930. *Procas armillatus* F. in abundance near Brighton. *Entomologist's monthly Magazine* **66**: 231.

CROWSON, R.A., 1967. *The Natural Classification of the Families of Coleoptera.* (Reprint). Classey, Hanworth.

CUPPEN, J.G.M. & HEIJERMAN, T., 1995. A description of the larva of *Bagous brevis* Gyllenhal, 1836 (Coleoptera: Curculionidae) with notes on its biology. *Elytron* **9**: 45-63.

DEJEAN, P.F.M.A., 1821. *Catalogue de la collection de Coléoptères de M. le Baron Dejean.* Crevot, Paris.

DIECKMANN, L., 1964. Die mitteleuropäischen Arten aus der Gattung *Bagous* Germ. *Entomologische Blätter für Biologie und Systematik der Käfer* **60**: 88-111.

DIECKMANN, L., 1981. Die *Hypera dauci*-Gruppe (Coleoptera, Curculionidae). *Reichenbachia* **19**: 111-116.

DIECKMANN, L., 1983. Beiträge zur Insektenfauna der DDR: Coleoptera - Curculionidae (Tanymecinae, Leptopiinae, Cleoninae, Tanyrhynchinae, Cossoninae, Raymondionyminae, Bagoinae, Tanysphyrinae). *Beiträge zur Entomologie* **33**: 257-381.

DIECKMANN, L., 1986. Beiträge zur Insektenfauna der DDR: Coleoptera - Curculionidae (Erirhinae). *Beiträge zur Entomologie* **36**: 119-181.

DIECKMANN, L., 1989. Die Zucht mitteleuropäischer Hyperini-Arten (Coleoptera, Curculionidae). *Entomologische Nachrichten und Berichte* **33**: 97-102.

DIECKMANN, L., 1990. Revision der mitteleuropäischen Arten der *Bagous collignensis*-Gruppe. *Reichenbachia* **27**: 141-145

DONISTHORPE, H.St.J.K., 1925. *Dryophthorus corticalis* Pk., a genus and species of Coleoptera new to Britain. *Entomologist's monthly Magazine* **61**: 182.

DONISTHORPE, H.St.J.K., 1931. *An Annotated List of the Additions to the British Coleopterous Fauna.* Nathaniel Lloyd, London.

DONISTHORPE, H.St.J.K., 1947. *Dorytomus filirostris* Gyll. (Col., Curculionidae) at Wicken Fen, Cambs. *Entomologist's monthly Magazine* **83**: 127.

DRANE, A.B., 1979. *Cossonus linearis* (F.) and *C. parallelepipedus* (Herbst) occurring together in a willow at Wicken Fen Nature Reserve, Cambs. *Entomologist's monthly Magazine* **114** (1978): 200.

DUFF, A., 1993. *Beetles of Somerset.* Somerset Archaeological & Natural History Society, Taunton.

DUFFY, E.A.J., 1954. Coleoptera. Scolytidae and Platypodidae. *Handbooks for the Identification of British Insects* **5** (15): 1-20.

DUMPER, E., 1937. *Cleonus albidus* F. (Col., Curculionidae) in the New Forest, Hants. *Journal of the Society for British Entomology* **1** (7): 172.

EVERSHAM, B.C. & TELFER, M.G., 1995. Coleoptera. *British Journal of Entomology and natural History* **8**: 202.

FLOYD, E.H. & NEWSOM, L.D., 1959. Biological study of the rice weevil complex. *Annals of the entomological Society of America* **52**: 687-695.

FOLWACZNY, B., 1973. Bestimmungstabelle der paläarktischen Cossoninae (Coleoptera, Curculionidae) ohne die nur in China und Japan vorkommenden Gattungen nebst Angaben zur Verbreitung. *Entomologische Blätter für Biologie und Systematik der Käfer* **69**: 65-180.

FOLWACZNY, B., 1983. Unterfamilie Cossoninae. pp. 30-43 *in* FREUDE, H., HARDE, K.W. & LOHSE, G.A. (Eds.) *Die Käfer Mitteleuropas* **11**. Goecke & Evers, Krefeld.

FOWLER, W.W., 1881. The Coleoptera of Askham Bog, York. *Entomologist's monthly Magazine* **18**: 7-9.

FOWLER, W.W., 1891. *The Coleoptera of the British Islands*. **5**. Reeve, London.

FOWLER, W.W. & DONISTHORPE, H.St.J.K., 1913. *The Coleoptera of the British Islands*. **6**. Reeve, Ashford.

FOWLES, A.P., 1992. Observations on *Procas granulicollis* Walton. *Coleopterist* **1**: 19-20.

FOWLES, A.P., 1994. A provisional key to the weevils of the genus *Hypera* (Germar) (Curculionidae). *Coleopterist* **3**: 15-20.

FOWLES, A.P. & HAMMETT, M.J., 2001. Confirmation of Rock Samphire *Crithmum maritimum* L. as a larval foodplant of the weevil *Hypera pollux* (Fabricius) (Curculionidae) in Britain. *Coleopterist* **10**: 27-28.

GONZALEZ, M., 1967. El género *Orthochaetes* Germar (Col. Curculionidae). *Publicaciones del Instituto de Biologia aplicada* **42**: 49-85.

GRATWICK, M. (Ed.), 1992. *Crop Pests in the UK. Collected edition of MAFF leaflets*. Chapman & Hall, London.

HALSTEAD, D.G.H., 1964. The separation of *Sitophilus oryzae* (L.) and *S. zeamais* Motschulsky (Col., Curculionidae), with a summary of their distribution. *Entomologist's monthly Magazine* **99** (1963) : 72-74.

HAMMAD, S.M., 1955. The immature stages of *Pentarthrum huttoni* Woll. (Coleoptera: Curculionidae). *Proceedings of the Royal entomological Society of London (A)* **30**: 33-39.

HAMMOND, P.M. 1998. *A taxonomic review of possibly endemic British non-marine invertebrates (with additional data)*. Unpublished Report (LOW/VT12H) to English Nature, Peterborough. 79 pp., 2 appendices. The Natural History Museum, London.

HARRIS, J.W.E. & COPPEL, H.C., 1967. The Poplar-and-Willow Borer, *Sternochetus* (=*Cryptorhynchus*) *lapathi* (Coleoptera: Curculionidae) in British Columbia. *Canadian Entomologist* **99**: 411-418.

HEAL, N.F., 1992. The discovery of *Lixus scabricollis* Bohe. (Curculionidae) in Britain. *Coleopterist* **1**: 2.

HEIJERMAN, T., 1999. The fern weevil, *Syagrius intrudens* Waterhouse (Curculionidae) on Guernsey. *Coleopterist* **8**: 38-39.

HERING, M., 1957. *Biology of the Leaf Miners*. Junk, 's-Gravenhage.

HEY, W.C., 1895. *Elmidomorphus aubei* (*Bagous petro*) at Askham Bog. *Naturalist* **20**: 242.

HICKIN, N.E., 1968. *The Insect Factor in Wood Decay* (second edition), Hutchinson, London.

HINTON, H.E., 1976. Plastron respiration in bugs and beetles. *Journal of Insect Physiology* **22**: 1529-1550.

HODGE, P. J. & JONES, R. A., 1995. *New British Beetles. Species not in Joy's practical handbook*. British Entomological and Natural History Society, Reading.

HOFFMANN, A., 1958. Coléoptères Curculionides (Troisième Partie). *Faune de France* **62**: 1209-1839.

HUM, M., GLASER, A.E & EDWARDS, R., 1980. Wood-boring weevils of economic importance in Britain. *Journal of the Institute of Wood Science* **22**: 201-207.

HYMAN, P.S. & PARSONS, M.S., 1992. *A review of the scarce and threatened Coleoptera of Great Britain. Pt. 1*. JNCC, Peterborough.

JANSON, O.E., 1921. *Stenopelmus rufinasus* Gyll., an addition to the list of British Coleoptera. *Entomologist's monthly Magazine* **57**: 225-226.

JOY, N.H., 1932. *A Practical Handbook of British Beetles* (2 vols.). Witherby, London.

KENWARD, H.K., 1990. A belated record of *Procas granulicollis* Walton (Col., Curculionidae) from Galloway, with a discussion of the British *Procas* spp. *Entomologist's monthly Magazine* **126**: 21-25.

KEYS, J.H., 1916. *Anchonidium unguiculare* Aube: a genus and species of Coleoptera new to the British list. *Entomologist's monthly Magazine* **52**: 112-113.

KIPPENBERG, H., 1983. Unterfamilie Hylobiinae. pp. 121-154 *in* FREUDE, H., HARDE, K.W. & LOHSE, G.A. (Eds.) *Die Käfer Mitteleuropas* **11**. Goecke & Evers, Krefeld.

KLOET, G.S. & HINCKS, W.D., 1945. *A check list of British Insects*. Kloet & Hincks, Stockport.

KUSCHEL, G., 1962. Some notes on the cossonine genus *Caulophilus* Wollaston with a key to the species (Coleoptera: Curculionidae). *Coleopterists' Bulletin* **16**: 1-4.

KUSCHEL, G., 1964. Insects of Campbell Island. Coleoptera: Curculionidae of the subantarctic islands of New Zealand. *Pacific Insects Monographs* **7**: 416-493.

KUSCHEL, G., 1987. The subfamily Molytinae (Coleoptera: Curculionidae): General notes and descriptions of new taxa from New Zealand and Chile. *New Zealand Entomologist* **9**: 11-29.

KUSCHEL, G., 1991. A pitfall trap for hypogean fauna. *Curculio* **31**: 5.

KUSCHEL, G., 1995. A phylogenetic classification of Curculionoidea to families and subfamilies. *Memoirs of the entomological Society of Washington* **14**: 5-33.

LAIDLAW, W.B.R., 1931. The pine-cone weevil (*Pissodes validirostris*) in Britain, with a brief comparative account of the genus *Pissodes*. *Scottish Naturalist* **1931**: 79-84.

LAWRENCE, J.F. & NEWTON, A.F., 1995. Families and subfamilies of Coleoptera (with selected genera, notes, references and data on family-group names). pp. 779-1006 *in* PAKALUK, J. & SLIPINSKI, S.A. (Eds.) *Biology, Phylogeny and Classification of Coleoptera*. Muzeum i Instytut Zoologii PAN, Warsaw.

LIOTTA, G., 1963. Osservazioni sul *Lixus algirus* L. (Punteruolo degli steli delle fave) (Col. Curculionidae). *Bollettino dell'Instituto di Entomologia agraria e dell'Osservatorio di Fitopatologia di Palermo* **5** (33): 1-24.

LOHSE, G.A.., 1983. Unterfamilie Pissodinae. pp. 110-120 *in* FREUDE, H., HARDE, K.W. & LOHSE, G.A. (Eds.) *Die Käfer Mitteleuropas* **11**. Goecke & Evers, Krefeld.

LUCHT, W.H., 1987. *Die Käfer Mitteleuropas: Katalog*. Goecke & Evers, Krefeld.

LYAL, C.H.C. & KING, T. 1996. Elytro-tergal stridulation in weevils (Insecta: Coleoptera: Curculionoidea). *Journal of natural History* **30**: 703-773.

MANGAN, J., 1908. The life-history of *Syagrius intrudens*, Waterh. a destructive fern-eating weevil. *Journal of economic Biology* **3**: 84-91.

MARSHALL, G.A.K., 1922. On the Australian fern weevils. *Bulletin of entomological Research* **13**: 169-174.

MASSEE, A.M., 1940. *Lixus paraplecticus* L. (Col., Curculionidae) in Kent. *Entomologist's monthly Magazine* **76**: 22.

MASSEE, A.M., 1954. *The Pests of Fruit and Hops*. (3rd ed.) Crosby Lockwood, London.

MASSEE, A.M., 1963. [no title] *Proceedings and Transactions of the South London entomological and natural History Society* **1962**: 5.

MAY, B.M., 1993. Larvae of Curculionoidea (Insecta: Coleoptera): a systematic overview. *Fauna of New Zealand* **28**: 1-226.

MEIKLE, R.D., 1984. *Willows and Poplars of Great Britain and Ireland*. Botanical Society of the British Isles, London.

MITFORD, R.S., 1923. *Lixus algirus* L. at Fairlight. *Entomologist's Record and Journal of Variation* **35**: 158.

MORIMOTO, K., 1962a. Comparative morphology and phylogeny of the superfamily Curculionoidea of Japan. (Comparative morphology, phylogeny and systematics of the superfamily Curculionoidea of Japan. I). *Journal of the Faculty of Agriculture, Kyushu University* **11**: 331-373.

MORIMOTO, K., 1962b. Descriptions of a new subfamily, new genera and species of the family Curculionidae of Japan. (Comparative morphology, phylogeny and systematics of the superfamily Curculionoidea of Japan. II). *Journal of the Faculty of Agriculture, Kyushu University* **11**: 374-409.

MORRIS, M.G., 1970. Notes on the life-history of *Dorytomus hirtipennis* Bedel (Col., Curculionidae). *Entomologist's monthly Magazine* **105** (1969): 207-209.

MORRIS, M.G., 1990. Orthocerous Weevils. Coleoptera: Curculionoidea (Nemonychidae, Anthribidae, Urodontidae, Attelabidae and Apionidae). *Handbooks for the Identification of British Insects* **5** (16): 1-108.

MORRIS, M.G., 1993a. A critical review of the weevils (Coleoptera, Curculionoidea) of Ireland and their distribution. *Biology and Environment: Proceedings of the Royal Irish Academy* **93B**: 69-84.

MORRIS, M.G., 1993b. Some species of weevils recorded erroneously from Ireland, and others whose status requires confirmation (Coleoptera: Curculionoidea). *Irish Naturalists' Journal* **24**: 325-328.

MORRIS, M.G., 1995a. Recent advances in the higher systematics of Curculionoidea, as they affect the British fauna. *Coleopterist* **4**: 21-30.

MORRIS, M.G., 1995b. An enquiry into the status and biology of *Hypera ononidis* (Chevrolat) (Col., Curculionidae). *Entomologist's monthly Magazine* **131**: 141-150.

MORRIS, M.G., 1997. Broad-nosed Weevils. Coleoptera: Curculionidae (Entiminae). *Handbooks for the Identification of British Insects* **5** (17a): 1-106.

MORRIS, M.G., 1998. Comparative aspects of the biology of three species of *Dorytomus* (Col., Curculionidae) associated with Aspen, *Populus tremula* L. *Entomologist's monthly Magazine* **134**: 197-213.

MORRIS, M.G., 1999. Some records of weevils (Curculionoidea) from Sutherland and Caithness, Northern Scotland. *Coleopterist* **8**: 57-62.

MORRIS, M.G., 2000. *Thryogenes fiorii* (Erirhinidae) at Woodwalton Fen, Cambridgeshire. *Coleopterist* **9** (2): 101.

MORRIS, M.G. & BOOTH, R.G., 1997. Notes on the nomenclature of some British weevils (Curculionoidea). *Coleopterist* **6** (3): 91-99.

MOUND, L. (ed.), 1989. Common insect pests of stored food products (seventh ed.). *British Museum (Natural History) Publications (Economic Series)* **15**: 1-68.

MURRAY, A., 1853. *Catalogue of the Coleoptera of Scotland.* Blackwood & Sons, Edinburgh.

NASH, D.R., 1979. *Cossonus linearis* (F.) and *C. parallelepipedue [sic]* (Herbst) inhabiting the same stump in the Suffolk Breck. *Entomologist's monthly Magazine* **114** (1978): 89.

NEWBERY, E.A., 1902. A revision of the British species of *Bagous* Schoen. *Entomologist's Record and Journal of Variation* **14**: 149-156.

NEWMAN, E., 1858. *Polyommatus artaxerxes* and *P. agestis. Zoologist* **16**: 6211-6212.

O'BRIEN, C.W., 1970. A taxonomic revision of the genus *Dorytomus* in North America (Coleoptera: Curculionidae). *University of California Publications on Entomology.* **60**: 1-80.

O'BRIEN, C.W. & WIBNER, G.J., 1982. Annotated checklist of the weevils (Curculionidae *sensu lato*) of North America, Central America, and the West Indies. *Memoirs of the American entomological Institute* **34**: i-ix, 1-382.

O'BRIEN, C.W. & WIBNER, G.J., 1984. Annotated checklist of the weevils (Curculionidae *sensu lato*) of North America, Central America, and the West Indies (Coleoptera: Curculionoidea) - Supplement 1. *Southwestern Entomologist* **9**: 286-307.

OEVERING, P., MATTHEWS, B. J., & PITMAN, A. J., 2000. The intertidal cossonid weevil *Pselactus spadix* (Herbst) (Curculionidae) in England and Wales. *Coleopterist* **9**: 9-13.

OSELLA, G., 1977. Revisione della sottofamiglia Raymondionyminae (Coleoptera, Curculionidae). *Memorie del Museo civico di Storia naturale di Verona (IIa serie) Sezione Scienze della Vita* **1**: 1-162.

OSELLA, G., 1979. Soil Curculionidae (Coleoptera). *Bolletino di Zoologia* **46**: 299-318.

OWEN, J.A., 1983. *Dryophthorus corticalis* Payk. (Col., Curculionidae) struggles to survive at Windsor. *Entomologist's monthly Magazine* **119**: 224.

OWEN, J.A., 1995. A pitfall trap for repetitive sampling of hypogean arthropod faunas. *Entomologist's Record and Journal of Variation* **107**: 225-228.

OWEN, J.A., 1997. Observations on *Raymondionymus marqueti* (Aubé) (Col: Curculionidae) in north Surrey. *Entomologist* **116**: 122-129.

PALM, E., 1999. Nye arter og landskabsfund for snudebiller (Coleoptera: Curculionidae) i Sverige. *Entomologisk Tidskrift* **120**: 143-147.

PAPP, C.S., 1979. *An illustrated catalog of the Cryptorhynchinae of the New World with generic descriptions, references to the literature and deposition of type material (Coleoptera: Curculionidae).* State of California Department of Food and Agriculture, Division of Plant Industry, Laboratory Services - Entomology, [California].

PETRI, K., 1901. Monographie des Coleopteren-Tribus Hyperini. *Bestimmungs-Tabellen der europäischen Coleopteren* **44**: 1-42.

POPE, R.D. (Ed.), 1977. A check list of British Insects. Second Edition (completely revised). Part 3: Coleoptera and Strepsiptera. *Handbooks for the Identification of British Insects* **11** (3): i-xiv, 1-105.

READ, R.W.J., 1976. Notes on the biology of *Cleopus pulchellus* Herbst (Coleoptera: Curculionidae). *Entomologist's Gazette* **27**: 118-122.

READ, R.W.J., 1977. Notes on the biology of *Cionus scrophulariae* (L.), together with preliminary observations on *C. tuberculosus* (Scopoli) and *C. alauda* (Herbst) (Col., Curculionidae). *Entomologist's Gazette* **28**: 183-202.

READ, R.W.J., 1982. *Mesites tardii* (Curtis) (Coleoptera: Curculionidae) new to West Cumbria, with notes on the species in Britain. *Entomologist's Gazette* **33**: 233-242.

READ, R.W.J., 1999. Two weevils (Curculionidae) new to Cumbria. *Coleopterist* **8**: 62.

RICHARDS, O.W., 1944. Two strains of the rice weevil, *Calandra oryzae* (L.) (Coleopt., Curculionidae). *Transactions of the Royal entomological Society of London* **94**: 197-200.

RICHERSON, P.J. & GRIGARICK, A.A., 1967. The life history of *Stenopelmus rufinasus* (Coleoptera: Curculionidae). *Annals of the entomological Society of America* **60**: 351-354.

RYE, E.C., 1865. Coleoptera. New British species, corrections of nomenclature, etc., noticed since the publication of the Entomologist's Annual, 1864. *Entomologist's Annual* **1865**: 37-80.

SAWYER, G.S. & CRAGG, S.M., 1995. Attack by the wood-boring weevil, *Pselactus spadix* on timbers in the intertidal and splash zones in ports in the U.K. *Material und Organismen* **29**: 67-79.

SCHERF, H., 1964. Die Entwicklungsstadien der mitteleuropäischen Curculioniden (Morphologie, Bionomie, Ökologie). *Abhandlungen der senckenbergischen naturforschenden Gesellschaft* **506**: 1-335.

SHIRT, D. (Ed.), 1987. *British Red Data Books: 2. Insects.* Nature Conservancy Council, [Peterborough].

SMITH, B.D. & STOTT, K.G., 1964. The life history and behaviour of the willow weevil *Cryptorhynchus lapathi* L. *Annals of applied Biology* **54**: 141-151.

SMRECZYNSKI, S., 1972. *Klucze do Oznaczania Owadów Polski. Czesc XIX Chrzaszcze - Coleoptera Zeszyt 98d Ryjkowce - Curculionidae Podrodzina Curculioninae.* Polskie Towarzystwo Entomologiczne, Warsaw.

STACE, C., 1991. *New Flora of the British Isles.* CUP, Cambridge.

STEPHENS, J.F., 1831. *Illustrations of British Entomology. Mandibulata.* **4**. Baldwin & Cradock, London.

TEMPÈRE, G., 1972. Nouvelles notes sur les Curculionidae de la faune française (Col.), Taxonomie, Chorologie, Ecologie, Ethologie. *Annales de la Société entomologique de France* (N.S.) **8**: 141-167.

TEMPÈRE, G., 1979. Sur divers *Leiosoma* de la faune française notamment des Pyrénées (Col. Curculionidae). *Nouvelle Revue d'Entomologie* **9**: 271-286.

TEMPÈRE, G. & PÉRICART, J., 1989. Coléoptères Curculionidae (Quatrième Partie: Compléments)). *Faune de France* **74**: 1-534.

THOMPSON, R.T., 1988. On the identity of *Cryptorhynchus harrisoni* Pool (Col., Curculionidae). *Entomologist's monthly Magazine* **124**: 165-166.

THOMPSON, R.T., 1989. A preliminary study of the weevil genus *Euophryum* Broun (Coleoptera: Curculionidae: Cossoninae). *New Zealand Journal of Zoology* **16**: 65-79.

THOMPSON, R.T., 1992. Observations on the morphology and classification of weevils (Coleoptera, Curculionidae) with a key to major groups. *Journal of natural History* **26**: 835-891.

THOMPSON, R.T., 1995. Raymondionymidae (Col., Curculionoidea) confirmed as British. *Entomologist's monthly Magazine* **131**: 61-64.

THOMPSON, R.T. & HOWELL, A. 2001. The Australian Fern Weevil *Syagrius intrudens* Waterhouse (Curculionidae) confirmed on Guernsey. *Coleopterist* **10** (2): 48-49.

WELCH, R. C., 1990. *Macrorhynchus littoralis* (Broun) (Col., Curculionidae) a littoral weevil new to the Palaearctic region, from two sites in Kent. *Entomologist's monthly Magazine* **126**: 97-101.

WILLIAMS, S.A., 1968. *Raymondionymus marqueti* typical form in Surrey. *Entomologist's monthly Magazine* **104**: 112.

WINGELMÜLLER, A., 1937. Monographie der paläarktischen Arten der Tribus Cionini. *Koleopterologische Rundschau* **23**: 143-221.

WOLLASTON, T.V., 1860. On certain musical Curculionidae. *Annals and Magazine of natural History* **6**: 14-19.

WOLLASTON, T.V., 1873. On the genera of the Cossonidae. *Transactions of the entomological Society of London* **1873**: 427-657.

ZHERIKHIN, V.V. & GRATSHEV, V.G., 1995. A comparative study of the hind wing venation of the superfamily Curculionoidea with phylogenetic implications. pp. 633-777 in PAKALUK, J. & SLIPINSKI, S.A. (Eds.) *Biology, Phylogeny and Classification of Coleoptera.* Muzeum i Instytut Zoologii PAN, Warsaw.

ZIMMERMAN, E.C., 1993. *Australian Weevils.* **3**: *Nanophyidae, Rhynchophoridae, Erirhinidae, Curculionidae: Amycterinae, Literature consulted.* CSIRO, Canberra.

ZIMMERMAN, E.C., 1994a. *Australian Weevils.* **1**: *Orthoceri Anthribidae to Attelabidae, The Primitive Weevils.* CSIRO, Canberra.

ZIMMERMAN, E.C., 1994b. *Australian Weevils.* **2**: *Brentidae. Eurhynchidae, Apionidae, and a Chapter on Immature Stages by Brenda May.* CSIRO, Canberra.

ZUMPT, F., 1936. Revision der paläarktischen Arten der Gattung *Lepyrus* Germ. *Pubblicazioni de Museo entomologico "Pietro Rossi" Duino* **14**: 259-290.

ZWÖLFER, H., 1965. Preliminary list of phytophagous insects attacking wild Cynareae (Compositae) in Europe. *Technical Bulletin of the Commonwealth Institute of Biological Control* **6**: 81-154.

ZWÖLFER, H. & HARRIS, P., 1984. Biology and host specificity of *Rhinocyllus conicus* (Froel.) (Col., Curculionidae), a successful agent for biological control of the thistle, *Carduus nutans* L. *Zeitschrift für angewandte Entomologie* **97**: 36-62.

Table 1.
Foodplants, with feeding sites, of weevils (confirmed or assumed larval hosts in bold; host records based on continental work marked by *

Weevil species	Plant hosts	Feeding site
RAYMONDIONYMINAE		
Ferreria marqueti	?introduced conifers	?roots
DRYOPHTHORINAE		
Dryophthorus corticalis	**dead wood**	
Sitophilus spp.	**stored products**	grain
ERIRHININAE		
Stenopelmus rufinasus	***Azolla filiculoides***	foliage
Procas armillatus	[unknown]	
Procas granulicollis	?*Ceratocapnos claviculata*	?
	?*Pteridium aquilinum*	?
Notaris acridulus	*Glyceria maxima**	stems
Notaris bimaculatus	*Phalaris arundinacea*	stems
	Phragmites australis	stems
	Typha latifolia	stems
	Carex spp.	stems
Notaris scirpi	*Carex* spp.	stems
	Typha spp.	stems
	Scirpus spp.*	stems
Erirhinus aethiops	*Sparganium erectum*	stems
	Carex spp.	stems
Thryogenes festucae	"*Scirpus*" spp. (cf. Stace, 1991)	stems
Thryogenes fiorii	***Carex elata***	stems
	*Carex paniculata**	stems
Thryogenes nereis	***Eleocharis palustris***	stems
Thryogenes schirrosus	***Sparganium erectum*** (and spp.)	stems
Grypus equiseti	***Equisetum arvense***	stems
	Equisetum palustre	stems
LIXINAE		
Coniocleonus nebulosus	***Calluna vulgaris***	stems
Coniocleonus hollbergi	?*Pinus* spp.*	?roots
	?*Rumex acetosella**	?
Bothynoderes affinis	Chenopodiaceae*	root galls
	Atriplex spp.*	root galls
	Beta spp.*	root galls
	Chenopodium spp.*	root galls
Cleonis pigra	***Cirsium arvense***	stems
	Carduus spp.	stems
	Cirsium spp.	stems
Lixus paraplecticus	***Sium latifolium***	stems
	Oenanthe aquatica	stems
	Anthriscus spp.*	stems
	Apium spp.*	stems
	Berula spp.*	stems
Lixus iridis	Apiaceae*	stems
Lixus vilis	***Erodium cicutarium***	stems
Lixus angustatus	***Carduus*** spp.	stems
	Cirsium spp.	stems
	*Malva sylvestris**	stems
	*Vicia faba**	stems
Lixus elongatus	*Carduus* spp.*	stems
	Cirsium spp.*	stems

Weevil species	Plant hosts	Feeding site
Lixus scabricollis	**Atriplex** spp.	stems
	Beta spp.	stems
Larinus planus	**Carduus** spp.	flower-heads
	Cirsium spp.	flower-heads
	?Carlina vulgaris	flower-heads
	?Centaurea spp.	flower-heads
Rhinocyllus conicus	**Carduus** spp.	flower-heads
	Cirsium spp.	flower-heads
HYPERINAE		
Hypera punctata	**Trifolium pratense**	foliage
	Trifolium repens	foliage
	Trifolium spp.	foliage
	?Medicago spp.	foliage
Hypera dauci	**Erodium cicutarium**	foliage
	?Erodium spp.*	foliage
	?Geranium spp.*	foliage
Hypera arundinis	**Sium latifolium**	foliage
	?Other Apiaceae [? adults only]	
Hypera pollux	**Apium** spp.	foliage
	Crithmum maritimum	foliage
	Oenanthe crocata	foliage
	Oenanthe spp.	foliage
	Peucedenum spp.	foliage
	Daucus spp.	foliage
Hypera rumicis	**Rumex (Rumex)** spp.	foliage
	Rumex (R.) crispus	foliage
	Rumex (R.) hydrolapathum	foliage
	Rumex (R.) obtusifolius	foliage
	Polygonum spp.*	foliage
	Oxyria spp.*	foliage
	Rheum spp.*	foliage
Hypera arator	**Caryophyllaceae**	foliage and flowers
	Cerastium spp.	foliage and flowers
	Lychnis spp.	foliage and flowers
	Silene spp.	foliage and flowers
	Spergula spp.	foliage and flowers
	Spergularia spp.	foliage and flowers
Hypera pastinacae	**Daucus carota**	umbels and foliage
	*Pastinaca sativa**	foliage
Hypera diversipunctata	**Cerastium arvense**	foliage
	Myosoton aquaticum	foliage
	Stellaria alsine	foliage
	Stellaria media	foliage
Hypera suspiciosa	**Lathyrus** spp.	foliage
	Melilotus spp.	foliage
	Vicia spp.	foliage
	Vicia cracca	foliage
	?Trifolium	foliage
Hypera plantaginis	**Lotus corniculatus**	foliage and flowers
	?Lotus spp.	foliage and flowers
Hypera fuscocinerea	**Medicago** spp.	foliage
	Medicago falcata	foliage
	Medicago sativa	foliage
	Medicago lupulina	foliage
	Melilotus spp.*	foliage
	Trifolium spp.*	foliage
	Vicia spp.*	foliage

Weevil species	Plant hosts	Feeding site
Hypera postica	**Medicago** spp.	foliage
	Medicago lupulina	foliage
	Medicago sativa	foliage
	Melilotus spp.	foliage
	Trifolium spp.	foliage
Hypera meles	**Trifolium** spp.	foliage
	Trifolium pratense	foliage
	Trifolium repens	foliage
	*Trifolium arvense**	foliage
	*Trifolium incarnatum**	foliage
	?Lotus spp.*	foliage
	Medicago spp.*	foliage
Hypera venusta	**Anthyllis vulneraria**	foliage
	Ulex minor	foliage
	Lotus spp.	foliage
	Onobrychis spp.	foliage
	Ulex spp.	foliage
	Vicia spp.	foliage
Hypera nigrirostris	**Trifolium pratense**	foliage
	Trifolium spp.*	foliage
Hypera ononidis	**Ononis repens**	foliage
	Ononis spp.	foliage
Limobius mixtus	**Erodium cicutarium**	foliage
Limobius borealis	**Geranium** spp.	foliage
	Geranium pratense	foliage
	Geranium robertianum	foliage
	Geranium sanguineum	foliage
	?Geranium sylvaticum	foliage
	*Geranium molle**	foliage
	*Geranium pusillum**	foliage
	*Geranium pyrenaicum**	foliage
	*Erodium cicutarium**	foliage
CIONINAE		
Cionus alauda	**Scrophularia auriculata**	foliage
	Scrophularia nodosa	foliage
	Verbascum spp.	foliage
Cionus scrophulariae	**Scrophularia auriculata**	foliage
	Scrophularia nodosa	foliage
	Scrophularia scorodonia	foliage
	Buddleja davidii	foliage
	Buddleja globosa	foliage
	Phygelius capensis	foliage
Cionus tuberculosus	**Scrophularia auriculata**	foliage
	Scrophularia nodosa	foliage
	?Verbascum spp.	foliage
Cionus hortulanus	**Scrophularia auriculata**	foliage
	Scrophularia nodosa	foliage
	Verbascum thapsus	foliage
	Buddleja spp.	foliage
	?Verbascum spp.	foliage
Cionus longicollis	**Verbascum** spp.	foliage
	Verbascum thapsus	foliage
Cionus nigritarsis	**Verbascum nigrum**	foliage
	Verbascum thapsus	foliage
Cleopus pulchellus	**Scrophularia auriculata**	foliage
	Scrophularia nodosa	foliage
	Verbascum thapsus	foliage
	Verbascum spp.	foliage

136

Weevil species	Plant hosts	Feeding site
MOLYTINAE		
Lepyrus capucinus	*Salix* spp.*	?roots
Hylobius transversovittatus	**Lythrum salicaria**	rootstocks
Hylobius abietis	**Pinus** spp.	stumps etc.
	Pinus contorta	stumps etc.
	Pinus sylvestris	stumps etc.
	?other conifers	
Liparus germanus	**Heracleum sphondylium**	rootstocks
	?other Apiaceae	
Liparus coronatus	**Anthriscus sylvestris**	rootstocks
	?other Apiaceae	
Leiosoma deflexum	**Anemone nemorosa**	roots and rhizomes
	Caltha palustris	?roots
	Ranunculus spp.	roots and rhizomes
	Ranunculus repens	roots and rhizomes
Leiosoma oblongulum	**Anemone nemorosa**	?roots
	Ranunculus spp.	?roots
Leiosoma troglodytes	?*Ranunculus* spp.	
Mitoplinthus caliginosus	[polyphagous]	
	Humulus lupulus	rootstocks
Anchonidium unguiculare	[unknown]	
Trachodes hispidus	[unknown]	?dead wood
Syagrius intrudens	**ferns**	stems
	Pteridium aquilinum	stems
Pissodes pini	**Pinus sylvestris**	dead branches
	Pinus spp.	dead branches
Pissodes validirostris	**Pinus sylvestris**	cones
	Pinus spp.	cones
Pissodes castaneus	**Pinus sylvestris**	dead twigs
	Pinus spp.	dead twigs
CYCLOMINAE		
Gronops lunatus	**Spergularia marina**	unknown
	Spergularia media	unknown
	Caryophyllaceae	
	?*Cerastium* spp.	
	Spergula arvensis	
Gronops inaequalis	**Chenopodiaceae**	unknown
	Atriplex prostrata	unknown
MESOPTILIINAE (= MAGDALIDINAE)		
Magdalis armigera	**Ulmus** spp.	dead branches etc.
	Ulmus procera	dead branches etc.
Magdalis carbonaria	**Betula** spp.	dead branches etc.
Magdalis memnonia	Pinus spp.	dead branches etc.
Magdalis duplicata	Pinus sylvestris	dead branches etc.
Magdalis phlegmatica	Pinus sylvestris	dead branches etc.
Magdalis ruficornis	woody Rosaceae	dead twigs etc.
	Crataegus spp.	dead twigs etc.
	Prunus spp.	dead twigs etc.
	Prunus domestica	dead twigs etc.
Magdalis cerasi	**Quercus** spp.	dead twigs etc.
	woody Rosaceae*	dead twigs etc.
Magdalis barbicornis	**woody Rosaceae**	dead twigs etc.
	Crataegus spp.	dead twigs etc.
	Malus spp.	dead twigs etc.
	Mespilus spp.	dead twigs etc.
	Prunus spp.	dead twigs etc.

Weevil species	Plant hosts	Feeding site
	Pyrus spp.	dead twigs etc.
	Sorbus spp.	dead twigs etc.
ANOPLINAE		
Anoplus plantaris	*Betula* spp.	leaf-mines
Anoplus roboris	*Alnus glutinosa*	leaf-mines
COSSONINAE		
Pentarthrum huttoni	dead wood, mainly structural	
Euophryum rufum	dead wood, mainly structural	
Euophryum confine	dead wood, general	
Pselactus spadix	dead wood, maritime	
Caulophilus oryzae	stored products, especially ginger	
Pseudophloeophagus aeneopiceus	dead wood, general	
Cossonus linearis	dead wood, especially *Salix* and *Populus*	
Cossonus parallelepipedus	dead wood, especially *Salix* and *Populus*	
Rhopalomesites tardyi	dead wood, general (broad-leaved)	
Rhyncolus ater	dead wood (mainly *Pinus*)	
Phloeophagus gracilis	dead wood, general (broad-leaved)	
Phloeophagus lignarius	dead wood, general (broad-leaved)	
Stereocorynes truncorum	dead wood, general (broad-leaved)	
	dead wood of *Fagus*.	
Macrorhyncolus littoralis	dead wood, ?general	
CRYPTORHYNCHINAE		
Cryptorhynchus lapathi	*Salix* spp.	stems
	Salix alba	stems
	Salix triandra	stems
	Salix viminalis	stems
	?*Alnus* spp.	
	?*Betula* spp.	
Acalles misellus	*Crataegus* spp.	?dead twigs
	Hedera helix	?dead twigs
Acalles roboris	?*Quercus* spp.	?dead twigs
Acalles ptinoides	?*Calluna vulgaris*	?dead twigs
	?*Quercus* spp.	?dead twigs
TANYSPHYRINAE		
Tanysphyrus lemnae	*Lemna* spp.	thalli
	Lemna minor(*)	thalli
	*Calla palustris**	?
BAGOINAE		
Bagous alismatis	*Alisma plantago-aquatica*	leaf-mines
	*Sagittaria sagittifolia**	
Bagous petro	*Utricularia* spp.	?
	Utricularia vulgaris	?
Bagous tubulus	*Aleopecurus aequalis**	
	*Glyceria fluitans**	
	*Glyceria plicata**	
Bagous argillaceus	[unknown]	
Bagous binodulus	*Stratiotes aloides*	leaves*
Bagous nodulosus	*Butomus umbellatus*	stems
Bagous limosus	*Potamogeton* spp.	?
Bagous czwalinae	?grasses	?
Bagous tempestivus	?polyphagous	
	Ranunculus spp.	?stems
Bagous subcarinatus	*Ceratophyllum submersum*	?
Bagous brevis	*Ranunculus flammula*	stems and foliage

138

Weevil species	Plant hosts	Feeding site
Bagous longitarsis	?*Myriophyllum* spp.	?
Bagous collignensis	*Myriophyllum* spp.*	?
Bagous frit	*Menyanthes trifoliata*	?
Bagous lutulosus	**Juncus** spp.	?
	Juncus bufonius	?
Bagous diglyptus	*Saxifraga granulata**	?
Bagous lutosus	*Sparganium erectum**	
Bagous puncticollis	?polyphagous*	
Bagous glabrirostris	?polyphagous*	
Bagous lutulentus	*Equisetum fluviatile*	stems
Bagous robustus	?*Alisma plantago-aquatica**	

DORYTOMINAE

Dorytomus hirtipennis	**Salix alba**	male and female catkins
	*Salix eleagnos**	
	*Salix fragilis**	
	*Salix viminalis**	
Dorytomus ictor	**Populus nigra** *s.l.*	catkins
Dorytomus longimanus	**Populus nigra** *s.l.*	catkins
	Populus x canadensis var. **serotina**	catkins
	*Populus alba**	
Dorytomus tremulae	**Populus alba**	shoots, catkins*
	Populus tremula	shoots, catkins*
	?*Populus x canescens*	
Dorytomus tortrix	**Populus tremula**	male catkins
Dorytomus melanophthalmus	**Salix** spp.	catkins
	Salix aurita	catkins
	Salix capraea	catkins
	Salix cinerea	catkins
	Salix repens	catkins
Dorytomus salicis	**Salix repens**	catkins
	*Salix alba**	
	*Salix aurita**	
	*Salix capraea**	
	*Salix cinerea**	
Dorytomus affinis	**Populus tremula**	female catkins
Dorytomus dejeani	**Populus tremula**	male and female catkins
Dorytomus taeniatus	**Salix** spp.	catkins
	Salix aurita	catkins
	Salix capraea	catkins
	Salix cinerea	
	Salix reticulata	shoots
Dorytomus salicinus	**Salix** spp.	?catkins
	Salix aurita	?catkins
	Salix capraea	?catkins
	Salix cinerea	?catkins
Dorytomus majalis	**Salix** spp.	catkins
	Salix aurita	catkins
	Salix capraea	catkins
	Salix cinerea	catkins
Dorytomus rufatus	**Salix** spp.	catkins
	Salix aurita	catkins
	Salix capraea	catkins

STOREINAE

Pachytychius haematocephalus	**Lotus corniculatus**	seeds
	Lotus spp.*	

Weevil species	Plant hosts	Feeding site
STYPHLINAE		
Pseudostyphlus pillumus	***Matricaria recutita***	capitula
	Achillea spp.*	
	Anthemis spp.*	
	Matricaria spp.*	
	Tripleurospermum spp.*	
Orthochaetes setiger	extensively polyphagous	leaf-mines
	Allium ursinum	leaf-mines
Orthochaetes insignis	extensively polyphagous	leaf-mines
SMICRONYCHINAE		
Smicronyx reichi	***Centaurium erythraea***	seedheads
	?*Centaurium* spp.	
	?*Blackstonia perfoliata*	
	?*Gentianella* spp.	
Smicronyx coecus	***Cuscuta epithymum***	stem-galls
	?*Cuscuta europaea*	
Smicronyx jungermanniae	***Cuscuta epithymum***	stem-galls
	?*Cuscuta* spp.	

Table 2.
Plants, or other substrates, with which weevils are associated (in alphabetical order). Species with confirmed or assumed larval foodplants are in in bold; host records based on continental work marked by *

Plant host	Weevil species	Feeding site
Achillea spp.*	*Pseudostyphlus pillumus*	
Aleopecurus aequalis*	*Bagous tubulus*	
Alisma plantago-aquatica	*Bagous alismatis*	leaf-mines
?*Alisma plantago-aquatica**	*Bagous robustus*	?
Allium ursinum	*Orthochaetes setiger*	leaf-mines
Alnus glutinosa	*Anoplus roboris*	leaf-mines
?*Alnus* spp.	*Cryptorhynchus lapathi*	
Anemone nemorosa	*Leiosoma deflexum*	roots and rhizomes
	Leiosoma oblongulum	?roots
Anthemis spp.*	*Pseudostyphlus pillumus*	
Anthriscus spp.*	*Lixus paraplecticus*	stems
Anthriscus sylvestris	*Liparus coronatus*	rootstocks
Anthyllis vulneraria	*Hypera venusta*	foliage
Apiaceae*	*Lixus iridis*	stems
Apiaceae, general	*Hypera arundinis*	?adults only
?Apiaceae, general	*Liparus coronatus*	
	Liparus germanus	
Apium spp.	*Hypera pollux*	foliage
Apium spp.*	*Lixus paraplecticus*	stems
Atriplex prostrata	*Gronops inaequalis*	?
Atriplex spp.*	*Bothynoderes affinis*	root galls
Atriplex spp.	*Lixus scabricollis*	stems
Azolla filiculoides	*Stenopelmus rufinasus*	foliage
Berula spp.*	*Lixus paraplecticus*	stems
Beta spp.	*Lixus scabricollis*	stems
Beta spp.*	*Bothynoderes affinis*	root galls
Betula spp.	*Magdalis carbonaria*	dead branches etc.
	Anoplus plantaris	leaf-mines
?*Betula* spp.	*Cryptorhynchus lapathi*	stems

Plant host	Weevil species	Feeding site
?*Blackstonia perfoliata*	*Smicronyx reichi*	
Buddleja davidii	*Cionus scrophulariae*	
Buddleja globosa	*Cionus scrophulariae*	foliage
Buddleja spp.	*Cionus hortulanus*	foliage
Butomus umbellatus	*Bagous nodulosus*	stems
*Calla palustris**	*Tanysphyrus lemnae*	?
?*Calluna vulgaris*	*Acalles ptinoides*	?dead twigs
Calluna vulgaris	*Coniocleonus nebulosus*	stems
Caltha palustris	*Leiosoma deflexum*	?roots
Carduus spp.	*Larinus planus*	flower-heads
	Rhinocyllus conicus	flower-heads
	Lixus angustatus	stems
Carduus spp.	*Cleonis pigra*	stems
Carduus spp.*	*Lixus elongatus*	stems
?*Carlina vulgaris*	*Larinus planus*	flower-heads
Carex elata	*Thryogenes fiorii*	stems
*Carex paniculata**	*Thryogenes fiorii*	stems
Carex spp.	*Erirhinus aethiops*	stems
	Notaris bimaculatus	stems
	Notaris scirpi	stems
Caryophyllaceae	*Hypera arator*	foliage and flowers
Caryophyllaceae	*Gronops lunatus*	?
?*Centaurea* spp.	*Larinus planus*	flower-heads
Centaurium erythraea	*Smicronyx reichi*	seedheads
?*Centaurium* spp.	*Smicronyx reichi*	
Cerastium arvense	*Hypera diversipunctata*	foliage
Cerastium spp.	*Hypera arator*	foliage and flowers
?*Cerastium* spp.	*Gronops lunatus*	?
?*Ceratocapnos claviculata*	*Procas granulicollis*	?
Ceratophyllum submersum	*Bagous subcarinatus*	?
Chenopodiaceae	*Gronops inaequalis*	?
Chenopodiaceae*	*Bothynoderes affinis*	rootgalls
Chenopodium spp.*	*Bothynoderes affinis*	root galls
Cirsium arvense	*Cleonis pigra*	stems
Cirsium spp.	*Larinus planus*	flower-heads
	Rhinocyllus conicus	flower-heads
	Lixus angustatus	stems
Cirsium spp.	*Cleonis pigra*	stems
Cirsium spp.*	*Lixus elongatus*	stems
?conifers, general	*Hylobius abietis*	
Crataegus spp.	*Magdalis barbicornis*	dead twigs etc.
	Magdalis ruficornis	dead twigs etc.
Crataegus spp.	*Acalles misellus*	?dead twigs
Crithmum maritimum	*Hypera pollux*	foliage
Cuscuta epithymum	*Smicronyx coecus*	stem-galls
	Smicronyx jungermanniae	stem-galls
?*Cuscuta europaea*	*Smicronyx coecus*	
?*Cuscuta* spp.	*Smicronyx jungermanniae*	
Daucus carota	*Hypera pastinacae*	umbels and foliage
Daucus spp.	*Hypera pollux*	foliage
dead wood	*Dryophthorus corticalis*	
dead wood of *Fagus*.	*Stereocorynes truncorum*	
dead wood, general	*Euophryum confine*	
	Pseudophloeophagus aeneopiceus	
dead wood, ?general	*Macrorhyncolus littoralis*	
dead wood, general, broad-leaved	*Phloeophagus gracilis*	

141

Plant host	Weevil species	Feeding site
	Phloeophagus lignarius	
	Rhopalomesites tardyi	
	Stereocorynes truncorum	
dead wood, mainly *Pinus*	*Rhyncolus ater*	
dead wood, mainly structural	*Pentarthrum huttoni*	
	Euophryum rufum	
dead wood, maritime	*Pselactus spadix*	
dead wood, esp. *Salix* and *Populus*	*Cossonus linearis*	
	Cossonus parallelepipedus	
Eleocharis palustris	*Thryogenes nereis*	stems
Equisetum arvense	*Grypus equiseti*	stems
Equisetum fluviatile	*Bagous lutulentus*	stems
Equisetum palustre	*Grypus equiseti*	stems
Erodium cicutarium	*Limobius mixtus*	foliage
	Lixus vilis	stems
	Hypera dauci	foliage
*Erodium cicutarium**	*Limobius borealis*	foliage
?Erodium spp.*	*Hypera dauci*	foliage
ferns	*Syagrius intrudens*	stems
?Gentianella spp.	*Smicronyx reichi*	
*Geranium molle**	*Limobius borealis*	foliage
Geranium pratense	*Limobius borealis*	foliage
*Geranium pusillum**	*Limobius borealis*	foliage
*Geranium pyrenaicum**	*Limobius borealis*	foliage
Geranium robertianum	*Limobius borealis*	foliage
Geranium sanguineum	*Limobius borealis*	foliage
?Geranium sylvaticum	*Limobius borealis*	foliage
Geranium spp.	*Limobius borealis*	foliage
?Geranium spp.*	*Hypera dauci*	foliage
*Glyceria fluitans**	*Bagous tubulus*	?
*Glyceria maxima**	*Notaris acridulus*	stems
*Glyceria plicata**	*Bagous tubulus*	?
?grasses	*Bagous czwalinae*	?
Hedera helix	*Acalles misellus*	?dead stems
Heracleum sphondylium	*Liparus germanus*	rootstocks
Humulus lupulus	*Mitoplinthus caliginosus*	rootstocks
?introduced conifers	*Ferreria marqueti*	?roots
Juncus bufonius	*Bagous lutulosus*	?
Juncus spp.	*Bagous lutulosus*	?
Lathyrus spp.	*Hypera suspiciosa*	foliage
Lemna minor	*Tanysphyrus lemnae*	thalli*
Lemna spp.	*Tanysphyrus lemnae*	thalli
Lotus corniculatus	*Hypera plantaginis*	foliage and flowers
	Pachytychius haematocephalus	seeds
Lotus spp.	*Hypera venusta*	foliage
Lotus spp.*	*Pachytychius haematocephalus*	seeds
?Lotus spp.*	*Hypera meles*	foliage
?Lotus spp.	*Hypera plantaginis*	foliage and flowers
Lychnis spp.	*Hypera arator*	foliage and flowers
Lythrum salicaria	*Hylobius transversovittatus*	rootstocks
Malus spp.	*Magdalis barbicornis*	dead twigs etc.

Plant host	Weevil species	Feeding site
*Malva sylvestris**	*Lixus angustatus*	stems
Matricaria recutita	*Pseudostyphlus pillumus*	capitula
Matricaria spp.*	*Pseudostyphlus pillumus*	
Medicago falcata	*Hypera fuscocinerea*	foliage
Medicago lupulina	*Hypera postica*	foliage
Medicago lupulina	*Hypera fuscocinerea*	foliage
Medicago sativa	*Hypera fuscocinerea*	foliage
	Hypera postica	foliage
Medicago spp.	*Hypera fuscocinerea*	foliage
	Hypera postica	foliage
Medicago spp.*	*Hypera meles*	foliage
?*Medicago* spp.	*Hypera punctata*	foliage
Melilotus spp.*	*Hypera fuscocinerea*	foliage
Melilotus spp.	*Hypera postica*	foliage
	Hypera suspiciosa	foliage
Menyanthes trifoliata	*Bagous frit*	?
Mespilus spp.	*Magdalis barbicornis*	dead twigs etc.
Myosoton aquaticum	*Hypera diversipunctata*	foliage
?*Myriophyllum* spp.	*Bagous longitarsus*	?
Myriophyllum spp.*	*Bagous collignensis*	?
Oenanthe aquatica	*Lixus paraplecticus*	stems
Oenanthe crocata	*Hypera pollux*	foliage
Oenanthe spp.	*Hypera pollux*	foliage
Onobrychis spp.	*Hypera venusta*	foliage
Ononis repens	*Hypera ononidis*	foliage
Ononis spp.	*Hypera ononidis*	foliage
Oxyria spp.*	*Hypera rumicis*	foliage
*Pastinaca sativa**	*Hypera pastinacae*	foliage
Peucedenum spp.	*Hypera pollux*	foliage
Phalaris arundinacea	*Notaris bimaculatus*	stems
Phragmites australis	*Notaris bimaculatus*	stems
Phygelius capensis	*Cionus scrophulariae*	foliage
Picea spp.	*Hylobius abietis*	
Pinus contorta	*Hylobius abietis*	stumps etc.
Pinus sylvestris	*Pissodes validirostris*	cones
	Pissodes pini	dead branches
	Magdalis duplicata	dead branches etc.
	Magdalis phlegmatica	dead branches etc.
	Pissodes castaneus	dead twigs and branches
	Hylobius abietis	stumps etc.
Pinus spp.	*Pissodes validirostris*	cones
	Pissodes pini	dead branches
	Magdalis memnonia	dead branches etc.
	Hylobius abietis	stumps etc.
	Pissodes castaneus	dead twigs and branches
?*Pinus* spp.*	*Coniocleonus hollbergi*	?roots
Polygonum spp.*	*Hypera rumicis*	foliage
?polyphagous*	*Bagous glabrirostris*	?
	Bagous puncticollis	?
?polyphagous	*Bagous tempestivus*	?
[polyphagous]	*Mitoplinthus caliginosus*	
polyphagous, extensive	*Orthochaetes insignis*	leaf-mines
	Orthochaetes setiger	leaf-mines
Populus alba	*Dorytomus tremulae*	shoots, catkins*
*Populus alba**	*Dorytomus longimanus*	
Populus x **canadensis** v. **serotina**	*Dorytomus longimanus*	catkins

Plant host	Weevil species	Feeding site
?*Populus* x *canescens*	*Dorytomus tremulae*	
Populus nigra s.l.	*Dorytomus ictor*	catkins
	Dorytomus longimanus	catkins
Populus tremula	*Dorytomus affinis*	female catkins
	Dorytomus tortrix	male catkins
	Dorytomus dejeani	male and female catkins
	Dorytomus tremulae	shoots, catkins*
Potamogeton spp.	*Bagous limosus*	?
Prunus domestica	*Magdalis ruficornis*	dead twigs etc.
Prunus spp.	*Magdalis barbicornis*	dead twigs etc.
	Magdalis ruficornis	dead twigs etc.
Pteridium aquilinum	*Syagrius intrudens*	stems
?*Pteridium aquilinum*	*Procas granulicollis*	?
Pyrus spp.	*Magdalis barbicornis*	dead twigs etc.
Quercus spp.	*Magdalis cerasi*	dead twigs etc.
?**Quercus** spp.	*Acalles ptinoides*	?dead twigs
	Acalles roboris	?dead twigs
Ranunculus flammula	*Bagous brevis*	stems and foliage
Ranunculus repens	*Leiosoma deflexum*	roots and rhizomes
Ranunculus spp.	*Leiosoma oblongulum*	?roots
	Leiosoma deflexum	roots and rhizomes
	Bagous tempestivus	stems
?*Ranunculus* spp.	*Leiosoma troglodytes*	?
Rheum spp.*	*Hypera rumicis*	foliage
Rosaceae, woody	*Magdalis barbicornis*	dead twigs etc.
Rosaceae, woody*	*Magdalis cerasi*	dead twigs etc.
Rosaceae, woody	*Magdalis ruficornis*	dead twigs etc.
?*Rumex acetosella**	*Coniocleonus hollbergi*	?
Rumex (R.) crispus	*Hypera rumicis*	foliage
Rumex (R.) hydrolapathum	*Hypera rumicis*	foliage
Rumex (R.) obtusifolius	*Hypera rumicis*	foliage
Rumex (Rumex) spp.	*Hypera rumicis*	foliage
*Sagittaria sagittifolia**	*Bagous alismatis*	
Salix alba	*Dorytomus hirtipennis*	male and female catkins
	Cryptorhynchus lapathi	stems
*Salix alba**	*Dorytomus salicis*	
Salix aurita	*Dorytomus majalis*	catkins
	Dorytomus melanophthalmus	catkins
	Dorytomus rufatus	catkins
	Dorytomus taeniatus	catkins
	Dorytomus salicinus	?catkins
*Salix aurita**	*Dorytomus salicis*	
Salix capraea	*Dorytomus majalis*	catkins
	Dorytomus melanophthalmus	catkins
	Dorytomus rufatus	catkins
	Dorytomus taeniatus	catkins
	Dorytomus salicinus	?catkins
*Salix capraea**	*Dorytomus salicis*	
Salix cinerea	*Dorytomus majalis*	catkins
	Dorytomus melanophthalmus	catkins
	Dorytomus salicinus	?catkins
	Dorytomus taeniatus	catkins
*Salix cinerea**	*Dorytomus salicis*	
*Salix eleagnos**	*Dorytomus hirtipennis*	
*Salix fragilis**	*Dorytomus hirtipennis*	

Plant host	Weevil species	Feeding site
Salix repens	*Dorytomus melanophthalmus*	catkins
	Dorytomus salicis	catkins
Salix reticulata	*Dorytomus taeniatus*	shoots
Salix triandra	*Cryptorhynchus lapathi*	stems
Salix viminalis	*Cryptorhynchus lapathi*	stems
Salix viminalis*	*Dorytomus hirtipennis*	
Salix spp.	*Dorytomus melanophthalmus*	catkins
	Dorytomus taeniatus	catkins
	Dorytomus majalis	catkins
	Dorytomus rufatus	catkins
	Dorytomus salicinus	catkins
	Cryptorhynchus lapathi	stems/shoots
Salix spp.*	*Lepyrus capucinus*	?roots
Saxifraga granulata*	*Bagous diglyptus*	?
"*Scirpus*" spp.* (cf Stace, 1991)	*Notaris scirpi*	stems
	Thryogenes festucae	stems
Scrophularia auriculata	*Cionus alauda*	foliage
	Cionus hortulanus	foliage
	Cionus scrophulariae	foliage
	Cionus tuberculosus	foliage
	Cleopus pulchellus	foliage
Scrophularia nodosa	*Cionus alauda*	foliage
	Cionus hortulanus	foliage
	Cionus scrophulariae	foliage
	Cleopus pulchellus	foliage
Scrophularia nodosa	*Cionus tuberculosus*	foliage
Scrophularia scorodonia	*Cionus scrophulariae*	foliage
Silene spp.	*Hypera arator*	foliage and flowers
Sium latifolium	*Hypera arundinis*	foliage
	Lixus paraplecticus	stems
Sorbus spp.	*Magdalis barbicornis*	dead twigs etc.
Sparganium erectum (and spp.)	*Thryogenes schirrosus*	stems
Sparganium erectum	*Erirhinus aethiops*	stems
Sparganium erectum*	*Bagous lutosus*	?
Spergula spp.	*Hypera arator*	foliage and flowers
Spergula arvensis	*Gronops lunatus*	?
Spergularia marina	*Gronops lunatus*	?
Spergularia media	*Gronops lunatus*	?
Spergularia spp.	*Hypera arator*	foliage and flowers
Stellaria alsine	*Hypera diversipunctata*	foliage
Stellaria media	*Hypera diversipunctata*	foliage
stored products, especially ginger	*Caulophilus oryzae*	
stored products, grain	*Sitophilus* spp.	
Stratiotes aloides	*Bagous binodulus*	leaves*
Trifolium arvense*	*Hypera meles*	foliage
Trifolium incarnatum*	*Hypera meles*	foliage
Trifolium pratense	*Hypera meles*	foliage
	Hypera nigrirostris	foliage
	Hypera punctata	foliage
Trifolium repens	*Hypera meles*	foliage
	Hypera punctata	foliage
Trifolium spp.	*Hypera punctata*	foliage
Trifolium spp.	*Hypera postica*	foliage
Trifolium spp.	*Hypera meles*	foliage
Trifolium spp.*	*Hypera fuscocinerea*	foliage
	Hypera nigrirostris	foliage
?*Trifolium* spp.	*Hypera suspiciosa*	foliage

Plant host	Weevil species	Feeding site
Tripleurospermum spp.*	*Pseudostyphlus pillumus*	
Typha latifolia	*Notaris bimaculatus*	stems
Typha spp.	*Notaris scirpi*	stems
Ulex minor	*Hypera venusta*	foliage
Ulex spp.	*Hypera venusta*	foliage
Ulmus spp.	*Magdalis armigera*	dead branches etc.
Ulmus procera	*Magdalis armigera*	dead branches etc.
[unknown; ?dead wood]	*Trachodes hispidus*	
[unknown]	*Anchonidium unguiculare*	
	Bagous argillaceus	
	Procas armillatus	
Utricularia vulgaris	*Bagous petro*	?
Utricularia spp.	*Bagous petro*	?
Verbascum nigrum	*Cionus nigritarsis*	foliage
Verbascum thapsus	*Cionus hortulanus*	foliage
	Cionus longicollis	foliage
	Cleopus pulchellus	foliage
Verbascum thapsus	*Cionus nigritarsis*	foliage
Verbascum spp.	*Cionus longicollis*	foliage
Verbascum spp.	*Cionus alauda*	foliage
	Cleopus pulchellus	foliage
?Verbascum spp.	*Cionus hortulanus*	foliage
	Cionus tuberculosus	foliage
Vicia cracca	*Hypera suspiciosa*	foliage
*Vicia faba**	*Lixus angustatus*	stems
Vicia spp.	*Hypera suspiciosa*	foliage
Vicia spp.	*Hypera venusta*	foliage
Vicia spp.*	*Hypera fuscocinerea*	foliage

Index

Roman entries are to checklist and main entry, **bold** signifies figures and *italics* Table 1

148